ELSEVIER'S DICTIONARY OF CLIMATOLOGY AND METEOROLOGY

ELSEVIER'S DICTIONARY OF CLIMATOLOGY AND METEOROLOGY

in
English, French, Spanish, Italian, Portuguese and German

compiled by

J. L. DE LUCCA
Sao Paulo, Brazil

ELSEVIER
Amsterdam – Lausanne – New York – Oxford – Shannon – Tokyo 1994

ELSEVIER SCIENCE B.V.
Sara Burgerhartstraat 25
P.O. Box 211, 1000 AE Amsterdam, The Netherlands

Library of Congress Cataloging-in-Publication Data

Lucca, J. L. de.
Elsevier's dictionary of climatology and meteorology : in English, French, Spanish, Italian, Portuguese and German / compiled by J.L. de Lucca.
p. cm.
ISBN 0-444-81532-5 (acid free paper)
1. Meteorology--Dictionaries--Polyglot. 2. Climatology--Dictionaries--Polyglot. 3. Dictionaries, Polyglot. I. Title.
QC854.L83 1994
551.5'03--dc20
94-6656
CIP

ISBN: 0-444-81532-5

This book is printed on acid-free paper.

Printed in The Netherlands

Preface

This dictionary contains 3094 terms in common use in all fields of meteorological science. Special attention has been given to application fields such as agrometeorology, climatology, hydrometeorology, environmental meteorology, aeronautical meteorology, marine meteorology and radiometeorology.

Meteorology is of growing importance nowadays in daily life: weather forecast on television and radio and in daily newspapers; forest fires, atmospheric disturbances and other atmospheric phenomena are controlled by means of meteorological satellites.

The terms, taken from textbooks, bulletins, journals, periodicals and glossaries in the field of climatology and meteorology, are given in English, French, Spanish, Italian, Portugese and German. The dictionary will be a tool to translate maps, tables, climatic studies, air weather reports, weather and sea bulletins and climatological and meteorological publications.

Any comments, critical or otherwise, would be greatly appreciated.

J.L. De Lucca

Abbreviations

AC	altocumulus
AOSB	Arctic Ocean Sciences Board
AWP	allied weather publication
AWS	air weather service
AWX	all weather aircraft
AWX(F)	all weather fighter
B	blue sky
BAPMON	background air pollution monitoring network
BC	sky partly clouded
BIOMASS	biological investigation of marine antarctic systems and stocks
BIOTROP	Regional Centre for Tropical Biology (Indonesia)
C	Celsius
C	cloudy
CACGP	Commission on Atmospheric Chemistry and Global Pollution
CAeM	Commission for Aeronautical Meteorology
CB	cumulonimbus
CC	cirrocumulus
CCAMLR	Commission for Conservation of Antarctic Marine Living Resources
CDM	Commission on Dynamic Meteorology
CEEMAT	Centre d'Étude et d'Expérimentation du Machinisme Agricole Tropical
CMAé	Commission de Météorologie Aéronautique
CMO	Caribbean Meteorologic Institute
CMP	centre météorologique principal
CMS	centre météorologique secondaire
CTFT	Centre Technique Forestier Tropical
CU	cumulus
D	drizzle
DEFORPA	Dépérissement de Forêts et Pollution Atmosphérique
DENMET	Departamento Nacional de Meteorologia do Ministério da Agricultura e Reforma Agrária
DGAP	Département de Génétique et d'Amélioration des Plantes
DMN	Direction de la Météorologie Nationale (France)
DMO	dependent meteorological office
DSA	Département Système Agraire
E	wet air (without precipitation)
ECOR	Engineering Commission on Ocean Resources

F	Fahrenheit
F	fog
G	gale
GARP	Global Atmospheric Research Programme
GERDAT	Département de Gestion, Recherche, Documentation et Appui Technique
GIPME	Global Investigation of Pollution in the Marine Environment
H	hail
HAPEX	Hydrological and Atmospheric Pilot Experiment
IAMAP	International Association of Meteorology and Atmospheric Physics
IAWPRC	International Association on Water Pollution Research and Control
ICACGP	International Commission on Atmospheric Chemistry and Global Pollution
ICARDA	International Centre for Agricultural Research in Dry Areas
ICCP	International Commission on Cloud Physics
ICMUA	International Commission on Meteorology of the Upper Atmosphere
ICPAE	International Commission on Planetary Atmospheres and their Evolution
ICPM	International Commission on Polar Meteorology
ICRISAT	International Crops Research Institute for Semi-Arid Tropics
ICSI	International Commission on Snow and Ice
ICTA	International Centre for Tropical Agriculture
ICWQ	International Commission on Water Quality
IEMVT	Institut d'Élevage et de Médecine Vétérinaire des Pays Tropicaux
IITA	International Institute of Tropical Agriculture
IMC	instrument meteorological conditions
INRA	Institut National de la Recherche Agronomique
INTECOL	International Association for Ecology
IPCC	Intergovernmental Panel on Climate Change
IRAT	Institut de Recherches Agronomiques Tropicales et des Cultures Vivrières
ISB	International Society of Biometeorology
ISCCP	International Satellite Cloud Climatology Project
ISLSCP	International Satellite Land Surface Climatology Project
ISSS	International Society of Soil Science
JP	precipitation in sight of ship or station
JSC	Joint Scientific Committee of the World Climate Research Programme
K	Kelvin
KQ	line squall
KS	storm of drifting snow
KZ	sand storm or dust storm
L	lightning
M	mist
MET	meteorology

MONEX	monsoon experiment
MONSEE	monitoring the sun earth environment
MOTNE	Meteorological Operational Telecommunications Network in Europe
MWO	Meteorological Watch Office
NS	nimbostratus
NWSC	national weather satellite center
O	overcast sky
OFM	Office Français de Météorologie
OWS	ocean weather station
PAMOY	Programme Atmosphère Moyenne
PIGB	Programme International Géosphère et Biosphère
PSFG	Permanent Service on Fluctuation of Glaciers
Q	squally weather
R	rain
R	Reaumur
RS	sleet (rain and snow together)
S	snow
SACAD	Systèmes Agraires Caribéens et Alternatives de Développement
SC	stratocumulus
SCAR	Scientific Committee on Antarctic Research
SCOMO	satellite collection of meteorological observations
SIG	systèmes d'information géographique
SMO	supplementary meteorological office
SMS	synchronous meteorological satellite
ST	stratus
STP-MET	solar-terrestrial physics - meteorology
T	thunder
TLR	thunderstorm with rain
TLS	thunderstorm with snow
TOGA	tropical oceans and global atmosphere
U	ugly, threatening sky
V	unusual visibility
VMC	visual meteorological conditions
W	dew
WB	weather bureau
WBAS	weather bureau airport station
WCAP	World Climate Applications Programme
WCIP	World Climate Impacts Programme
WCDP	World Climate Data Programme
WCP	World Climate Programme
WCRP	World Climate Research Programme
WDC	World Data Centre
WITS	world ionosphere thermosphere study
WMO	World Meteorological Organization
WOCE	World Ocean Circulation Experiment

WWW	World Weather Watch
X	hoar frost
Y	dry air
Z	dust haze

Table 1 - Conversion of temperature

General formula: $K = C + 273$; $F = 9C : 5 + 32$

Freezing point of water: 0 C, 273 K, 32 F
Boiling point of water: 100 C, 373 K, 212 F

C	K	F	C	K	F	C	K	F	C	K	F
−50	223	−58	−12	261	11	26	299	78	64	337	147
−49	224	−56	−11	262	13	27	300	80	65	338	149
−48	225	−54	−10	263	14	28	301	82	66	339	150
−47	226	−52	−9	264	16	29	302	84	67	340	152
−46	227	−50	−8	265	18	30	303	86	68	341	154
−45	228	−49	−7	266	20	31	304	87	69	342	156
−44	229	−47	−6	267	22	32	305	89	70	343	158
−43	230	−45	−5	268	23	33	306	91	71	344	159
−42	231	−43	−4	269	25	34	307	93	72	345	161
−41	232	−41	−3	270	27	35	308	95	73	346	163
−40	233	−40	−2	271	29	36	309	96	74	347	165
−39	234	−38	−1	272	31	37	310	98	75	348	167
−38	235	−36	0	273	32	38	311	100	76	349	168
−37	236	−34	1	274	33	39	312	102	77	350	170
−36	237	−32	2	275	35	40	313	104	78	351	172
−35	238	−31	3	276	37	41	314	105	79	352	174
−34	239	−29	4	277	39	42	315	107	80	353	176
−33	240	−27	5	278	41	43	316	109	81	354	177
−32	241	−25	6	279	42	44	317	111	82	355	179
−31	242	−23	7	280	44	45	318	113	83	356	181
−30	243	−22	8	281	46	46	319	114	84	357	183
−29	244	−20	9	282	48	47	320	116	85	358	185
−28	245	−18	10	283	50	48	321	118	86	359	186
−27	246	−16	11	284	51	49	322	120	87	360	188
−26	247	−14	12	285	53	50	323	122	88	361	190
−25	248	−13	13	286	55	51	324	123	89	362	192
−24	249	−11	14	287	57	52	325	125	90	363	194
−23	250	−9	15	288	59	53	326	127	91	364	195
−22	251	−7	16	289	60	54	327	129	92	365	197
−21	252	−5	17	290	62	55	328	131	93	366	199
−20	253	−4	18	291	64	56	329	132	94	367	201
−19	254	−2	19	292	66	57	330	134	95	368	203
−18	255	0	20	293	68	58	331	136	96	369	204
−17	256	2	21	294	69	59	332	138	97	370	206
−16	257	4	22	295	71	60	333	140	98	371	208
−15	258	5	23	296	73	61	334	141	99	372	210
−14	259	7	24	297	75	62	335	143	100	373	212
−13	260	9	25	298	77	63	336	145			

Table 2 - Wind Chills

This table shows heat loss caused by a combination of temperature and wind on body surfaces. For example, a temperature of 30 degrees Fahrenheit, plus a wind of 45 miles per hour, causes a body heat loss which is equal to that in −6 degrees with no wind.

Fahrenheit	35	30	25	20	15	10	5	0	−5	−10	−15	−20	−25	−30	−35
MPH															
5	33	27	21	19	12	7	0	−5	−10	−15	−21	−26	−31	−36	−42
10	22	16	10	3	−3	−9	−15	−22	−27	−34	−40	−46	−52	−58	−64
15	16	9	2	−5	−11	−18	−25	−31	−38	−45	−51	−58	−65	−72	−78
20	12	4	−3	−10	−17	−24	−31	−39	−46	−53	−60	−67	−74	−81	−88
25	8	1	−7	−15	−22	−29	−36	−44	−51	−59	−66	−74	−81	−88	−96
30	6	−2	−10	−18	−25	−33	−41	−49	−56	−64	−71	−79	−86	−93	−101
35	4	−4	−12	−20	−27	−35	−43	−52	−58	−67	−74	−82	−89	−97	−105
40	3	−5	−13	−21	−29	−37	−45	−53	−60	−69	−76	−84	−92	−100	−107
45	2	−6	−14	−22	−30	−38	−46	−54	−62	−70	−78	−85	−93	−102	−109

(Source: The World Almanac and Book of Facts, 1980)

Basic Table

1 ablation
f ablation *f*
e ablación *f*
i ablazione *f*
p ablação *f*
d Ablation *f*

2 ablation area
f zone *f* d'ablation
e zona *f* de ablación
i zona *f* di ablazione
p zona *f* de ablação
d Ablationsgebiet *n*

3 ablation moraine
f moraine *f* d'ablation
e morrena *f* de ablación
i morena *f* di ablazione
p morena *f* de ablação
d Ablationsmoräne *f*

4 abnormal *adj*
f anormal
e anormal
i anormale
p anormal
d anormal

5 absolute humidity
f humidité *f* absolue
e humedad *f* absoluta
i umidità *f* assoluta
p umidade *f* absoluta
d absolute Feuchtigkeit *f*

6 absolute temperature
f température *f* absolue
e temperatura *f* absoluta
i temperatura *f* assoluta
p temperatura *f* absoluta
d absolute Temperatur *f*

7 absolute zero
f zéro *m* absolu
e cero *m* absoluto
i zero *m* assoluto
p zero *m* absoluto
d absoluter Nullpunkt *m*

8 absorption capacity
f pouvoir *m* absorbant
e poder *m* de absorción
i capacità *f* di assorbimento
p capacidade *f* de absorção
d Absorptionsvermögen *n*

9 absorption hygrometer; chemical hygrometer
f hygromètre *m* d'absorption; hygromètre *m* chimique
e higrómetro *m* de absorción; higrómetro *m* químico
i igrometro *m* di assorbimento
p higrômetro *m* de absorção; higrômetro *m* químico
d Absorptionshygrometer *n*

10 acceleration due to gravity
f accélération *f* de la pesanteur
e aceleración *f* de la gravedad
i accelerazione *f* di gravità
p aceleração *f* da gravidade
d Schwerebeschleunigung *f*

11 acclimatization
f acclimatation *f*
e aclimatación *f*
i acclimazione *f*; acclimatazione *f*
p aclimatação *f*
d Akklimatisierung *f*

12 acclimatization fever
f fièvre *f* d'acclimatation
e fiebre *f* de aclimatación
i febbre *f* di acclimatazione
p febre *f* de aclimatação
d Akklimatisationsfieber *n*

13 accretion
f accrétion *f*
e acreción *f*
i accrescimento *m*
p acreção *f*
d Wachstum *n*

14 accumulated temperature
f température *f* accumulée
e temperatura *f* acumulada
i temperatura *f* accumulata
p temperatura *f* acumulada
d Wärmesumme *f*

15 accumulation
f accumulation *f*
e acumulación *f*
i accumulo *m*
p acumulação *f*
d Anhäufung *f*

16 accumulation area
f zone *f* d'accumulation
e área *f* de acumulación
i area *f* di accumulo
p área *f* de acumulação
d Akkumulationsgebiet *n*

17 accuracy
f exactitude *f*
e exactitud *f*
i accuratezza *f*
p exatidão *f*
d Genauigkeit *f*

18 acid rain
f pluie *f* acide
e lluvia *f* ácida
i pioggia *f* acida
p chuva *f* ácida
d saurer Regen *m*

19 acid soil
f sol *m* acide
e suelo *m* ácido
i terreno *m* acido
p solo *m* ácido
d saurer Boden *m*

20 acrocyanosis
f acrocyanose *f*
e acrocianosis *f*
i acrocianosi *f*
p acrocianose *f*
d Akrozyanose *f*

21 actinic rays
f rayons *mpl* actiniques
e rayos *mpl* actínicos
i raggi *mpl* attinici
p raios *mpl* actínicos
d aktinische Strahlen *mpl*

22 actinograph
f actinographe *m*
e actinógrafo *m*
i attinografo *m*
p actinógrafo *m*; pireliógrafo *m*
d Aktinograph *m*

23 actinometer
f actinomètre *m*
e actinómetro *m*
i attinometro *m*
p actinômetro *m*
d Aktinometer *n*

24 actinometry
f actinométrie *f*
e actinometría *f*
i attinometria *f*
p actinometria *f*
d Aktinometrie *f*

25 action of atmospheric agents
f action *f* des agents atmosphériques
e acción *f* de los agentes atmosféricos
i azione *f* degli agenti atmosferici
p ação *f* dos agentes atmosféricos
d Einwirkung *f* der Atmosphärilien

26 action of ice
f action *f* des glaces
e acción *f* de los hielos
i azione *f* dei ghiacci
p ação *f* dos gelos
d Einwirkung *f* des Eises

27 action of snow
f action *f* de la neige
e acción *f* de la nieve
i azione *f* della neve
p ação *f* da neve
d Schnee(ein)wirkung *f*

28 active volcano
f volcan *m* actif
e volcán *m* activo
i vulcano *m* attivo
p vulcão *m* ativo
d tätiger Vulkan *m*

29 adaptation
f adaptation *f*
e adaptación *f*
i adattamento *m*
p adaptação *f*
d Anpassung *f*; Adaptation *f*

30 additional lighting
f éclairage *m* supplémentaire
e iluminación *f* suplementaria
i illuminazione *f* faggiuntiva
p iluminação *f* adicional
d Zusatzbelichtung *f*

31 adiabatic *adj*
f adiabatique
e adiabático
i adiabatico
p adiabático
d adiabatisch

32 adiabatic atmosphere
f atmosphère *f* adiabatique
e atmósfera *f* adiabática
i atmosfera *f* adiabatica
p atmosfera *f* adiabática
d adiabatische Atmosphäre *f*

33 adiabatic equilibrium; convective equilibrium
f équilibre *m* adiabatique; équilibre *m* convectif
e equilibrio *m* adiabático
i equilibrio *m* adiabatico; equilibrio *m* convettivo
p equilíbrio *m* adiabático; equilíbrio *m* convectivo
d adiabatisches Gleichgewicht *n*

34 adiabatic flow
f flux *m* adiabatique; écoulement *m* adiabatique
e corriente *f* adiabática
i flusso *m* adiabatico; corrente *f* adiabatica
p fluxo *m* adiabático
d adiabatische Strömung *f*

35 adiabatic lapse rate; dry-adiabatic lapse rate
f gradient *m* adiabatique
e gradiente *m* adiabático vertical
i gradiente *m* verticale adiabatico secco
p gradiente *m* vertical adiabático
d dynamische Erwärmung *f* und Abkühlung *f*

36 adiabatic pressure drop
f détente *f* adiabatique
e disminución *f* de presión adiabática
i caduta *f* di pressione adiabatica
p descida *f* de pressão adiabática
d adiabatischer Druckabfall *m*

37 adiabatic process
f processus *m* adiabatique
e proceso *m* adiabático
i trasformazione *f* adiabatica
p processo *m* adiabático
d adiabatisches Verfahren *n*

38 adsorption
f adsorption *f*
e adsorción *f*
i adsorbimento *m*
p adsorção *f*
d Adsorption *f*

39 advection
f advection *f*
e advección *f*
i avvezione *f*
p advecção *f*
d Advektion *f*

40 advection fog
f brouillard *m* d'advection
e niebla *f* de advección
i nebbia *f* d'avvezione
p nevoeiro *m* de advecção
d Advektionsnebel *m*

* **aeolian** *adj* → **947**

41 aeration; airing
f ventilation *f*
e ventilación *f*
i aerazione *f*
p aeração *f*
d Belüftung *f*

42 aeriform *adj*
f aériforme
e aeriforme
i aeriforme
p aeriforme
d gasförmig

43 aerobiosis
f aérobiose *f*
e aerobiosis *f*
i aerobiosi *f*
p aerobiose *f*
d Aerobiose *f*

44 aero-embolism
f aéroembolisme *m*
e aeroembolismo *m*
i embolia *f* gassosa
p aeroembolismo *m*
d Luftembolie *f*

45 aerogram
f aérogramme *m*
e aerograma *m*
i aerogramma *m*
p aerograma *m*
d Luftpostleichtbrief *m*

46 aerolite
f aérolithe *m*
e aerolito *m*
i aerolito *m*
p aerólito *m*
d Aerolith *m*

* **aerologation** → **2133**

47 aerology
f aérologie *f*
e aerología *f*
i aerologia *f*
p aerologia *f*
d Aerologie *f*; Luftkunde *f*

48 aerometeorograph
f aérométéorographe *m*
e aerometeorógrafo *m*
i aerometeorografo *m*
p aerometeorógrafo *m*
d Aerometeorograph *m*

49 aerometer
f aéromètre *m*
e aerómetro *m*
i aerometro *m*
p aerômetro *m*
d Aerometer *n*

50 aerometry
f aérométrie *f*
e aerometría *f*
i aerometria *f*
p aerometria *f*
d Aerometrie *f*

51 aeronautical meteorological station
f station *f* météorologique aéronautique
e estación *f* meteorológica aeronáutica
i stazione *f* meteorologica aeronautica
p estação *f* meteorológica aeronáutica
d Luftwetteramt *n*

52 aeroplankton
f aéroplancton *m*
e aeroplanctón *m*
i aeroplancton *m*
p aeroplâncton *m*
d Aeroplankton *n*

53 aeroscope
f aéroscope *m*
e aeroscopio *m*
i aeroscopio *m*
p aeroscópio *m*
d Aeroskop *n*

54 aeroscopy
f aéroscopie *f*
e aeroscopia *f*
i aeroscopia *f*
p aeroscopia *f*
d Aeroskopie *f*

55 aerosol
f aérosol *m*
e aerosol *m*
i aerosol *m*
p aerosol *m*
d Aerosol *n*

56 aerosphere
f aérosphère *f*
e aerosfera *f*
i aerosfera *f*
p aerosfera *f*
d Lufthülle *f*

57 aerostatics
f aérostatique *f*
e aerostática *f*
i aerostatica *f*
p aerostática *f*
d Aerostatik *f*

58 aerotropism
f aérotropisme *m*
e aerotropismo *m*
i aerotropismo *m*
p aerotropismo *m*
d Aerotropismus *m*

59 affluent
f affluent *m*
e afluente *m*
i affluente *m*
p afluente *m*
d Nebenfluß *m*

60 afterglow
f luminescence *f* résiduelle
e luminiscencia *f* residual
i luminescenza *f* residua
p luminescência *f* residual
d Nachglühen *n (Abendrot)*

61 afterglow time
f durée *f* de la persistance
e duración *f* de persistencia
i durata *f* della persistenza
p duração *f* da persistência
d Nachleuchtdauer *f*

62 afternoon; p.m.
f après-midi *m,f*
e tarde *f*
i pomeriggio *m*
p tarde *f*
d Nachmittag *m*

63 age of moon
f âge *m* de la lune
e edad *f* de la luna
i età *f* della luna
p idade *f* da lua
d Alter *n* des Mondes

64 age of tide
f âge *m* de la marée
e edad *f* de la marea
i età *f* della marea
p idade *f* da maré
d Alter *n* der Gezeiten

65 aggradation
f alluvionnement *m*
e agradación *f*
i sovralluvionamento *m*
p assoreamento *m*
d Anschwemmung *f*

66 agricultural climatology; agroclimatology
f climatologie *f* agricole
e climatología *f* agrícola
i climatologia *f* agraria
p climatologia *f* agrícola
d Agrarklimatologie *f*

67 agricultural meteorology; agrometeorology
f météorologie *f* agricole
e meteorología *f* agrícola
i meteorologia *f* agraria
p agrometeorologia *f*
d Agrarmeteorologie *f*

68 agriculture; farming
f agriculture *f*
e agricultura *f*
i agricoltura *f*; coltivazione *f*
p agricultura *f*
d Feldbau *m*

* **agroclimatology → 66**

* **agrometeorology → 67**

69 agronomy
f agronomie *f*; science *f* agronomique
e agronomía *f*
i agronomia *f*; scienza *f* agraria
p agronomia *f*
d Agronomie *f*; Agrarwissenschaft *f*

70 Agulhas Current
f Courant *m* des Agulhas
e Corriente *f* de las Agulhas
i Corrente *f* dei Agulhas
p Corrente *f* das Agulhas
d Agulhasstrom *m*

71 air
f air *m*
e aire *m*
i aria *f*
p ar *m*
d Luft *f*

72 airborne contaminants
f contaminants *mpl* dans l'air
e contaminantes *mpl* aerotransportados
i contaminanti *mpl* aerei
p poluentes *mpl* transportados através do ar
d in der Luft vorhandene aktive Teilchen *npl*

73 airborne particulates
f macroparticules *fpl* en suspension dans l'air
e macropartículas *fpl* en suspensión en al aire
i macroparticelle *fpl* in sospensione nell'aria
p micropartículas *fpl* en suspensão no ar
d mit der Luft mitgeführte Makroteilchen *npl*

74 air circulation fan
f ventilateur *m* de brassage
e ventilador *m* de circulación
i ventilatore *m* miscelatore
p ventoinha *f* axial
d Umwälzventilator *m*

75 air conditioning
f conditionnement *m* de l'air
e acondicionamiento *m* del aire
i condizionamento *m* d'aria
p condicionamento *m* do ar
d Klimatisierung *f*

76 air-conditioning humidification
f humectation *f* climatisante
e aspersión *f* de climatización
i irrigazione *f* climatizzante
p umidificação *f* de climatização
d klimatisierende Beregnung *f*

77 air-conditioning unit
f appareil *m* de climatisation
e aparato *m* climatizador de aire
i condizionatore *m* climatico
p unidade *f* de climatização
d Klimaanlage *f*

78 air contamination indicator
f signaleur *m* atmosphérique
e indicador *m* atmosférico
i segnalatore *m* atmosferico
p indicador *m* de poluição do ar
d Luftkontaminationsanzeiger *m*

79 air contamination meter
f contaminamètre *m* atmosphérique
e medidor *m* de contaminación atmosférica
i contaminametro *m* atmosferico
p medidor *m* de poluição atmosférica
d Gerät *n* zur Bestimmung der Luftkontamination

80 air contamination monitor
f moniteur *m* atmosphérique
e monitor *m* atmosférico
i monitore *m* atmosferico
p monitor *m* de poluição atmosférica
d Luftkontaminationswarngerät *n*

81 air content
f teneur *f* en air
e contenido *m* del aire
i contenuto *m* d'aria
p conteúdo *m* de ar
d Luftgehalt *m*

82 aircraft icing
f givrage *m* d'aéronef
e congelación *f* del avión
i formazione *f* di ghiaccio su un aereo
p formação *f* de gelo no avião
d Luftfahrzeugseisanlagerung *f*

83 air current
f courant *m* d'air
e corriente *f* de aire
i corrente *f* d'aria
p corrente *f* de ar
d Luftströmung *f*

84 air density
f densité *f* de l'air; masse *f* volumique de l'air
e densidad *f* del aire
i densità *f* dell'aria
p densidade *f* do ar
d Luftdichte *f*

*** airflow → 134**

85 air fog signal
f signal *m* de brume aérien
e señal *f* de niebla aérea
i segnale *m* di brume aerea
p sinal *m* de névoa aérea
d Luftnebelsignal *n*

*** airglow height → 1348**

86 airglow layer
f couche *f* émettrice
e estrato *m* resplandor
i strato *m* emittente
p estrato *m* luminescente
d Emissionsschicht *f*

87 air heater
f générateur *m* d'air chaud
e quemador *m* de aire caliente
i generatore *m* d'aria calda
p gerador *m* de ar quente
d Warmluftofen *m*

*** air humidity → 253**

*** airing → 41**

88 air mass
f masse *f* d'air
e masa *f* de aire
i massa *f* d'aria
p massa *f* de ar
d Luftmasse *f*

89 air medicine
f médecine *f* de l'aviation
e medicina *f* aeronáutica
i medicina *f* aeronautica
p medicina *f* aeronáutica
d Luftfahrtmedizin *f*

*** air moisture → 253**

*** air permeated with dust → 865**

90 air pocket
f trou *m* d'air
e pozo *m* de aire
i vuoto *m* d'aria
p poço *m* aéreo
d Luftloch *n*

91 air pollution; atmospheric pollution
f pollution *f* atmosphérique
e contaminación *f* atmosférica
i inquinamento *m* atmosferico
p poluição *f* atmosférica
d Luftverunreinigung *f*

92 air recirculation system
f système *m* de recirculation d'air
e sistema *m* de recirculación del aire
i sistema *m* di circolazione d'aria
p sistema *m* de recirculação de ar
d Umluftanlage *f*

93 air saturated with moisture
f air *m* saturé d'humidité
e aire *m* saturado de humedad
i aria *f* satura d'umidità
p ar *m* saturado de umidade
d mit Feuchtigkeit gesättigte Luft *f*

94 air speed indicator
f anémomètre *m* compensé
e indicador *m* de velocidad
i indicatore *m* de velocità
p indicador *m* de velocidade
d Geschwindigkeitsmesser *m*

95 air speed indicator calibrator
f vérificateur *m* d'installation anémométrique
e corrector *m* de anemómetro
i correttore *m* dell'anemometro
p corretor *m* de anemômetro
d Fahrtmessereichgerät *m*

96 air temperature
f température *f* de l'air
e temperatura *f* del aire
i temperatura *f* dell'aria
p temperatura *f* do ar
d Lufttemperatur *f*

97 air trajectory
f trajectoire *f* de l'air
e trayectoria *f* del aire
i traiettoria *f* dell'aria
p trajetória *f* do ar
d Luftbahn *f*

98 air want
f manque *m* d'air
e falta *f* de aire
i mancanza *f* d'aria
p falta *f* de ar
d Luftmangel *m*

99 Alaska Current
f Courant *m* du Alaska
e Corriente *f* del Alaska
i Corrente *f* dell'Alaska
p Corrente *f* do Alaska
d Alaskastrom *m*

100 albedo
(a measure of a reflective power of a surface or body)
f albédo *m*
e albedo *m*
i albedo *m*
p albedo *m*
d Albedo *m*

101 alcohol thermometer; spirit thermometer
f thermomètre *m* à alcool
e termómetro *m* de alcohol
i termometro *m* ad alcool
p termômetro *m* de álcool
d Alkoholthermometer *n*

102 Aleutian low
f dépression *f* des Aléoutes
e centro *m* de baja presión procedente de las Aleutas
i bassa *f* delle Aleutine
p ciclone *m* das Aleutas
d Aleuten-Tiefdruckgebiet *n*

103 alimentation
f alimentation *f*
e alimentación *f*
i alimentazione *f*
p alimentação *f*
d Ernärung *f*

104 allobar
(a change in barometric pressure)
f allobare *f*
e alobara *f*
i allobara *f*
p alóbara *f*
d Allobare *f*

105 allowed region
f région *f* permise
e región *f* permitida
i regione *f* permessa
p região *f* permitida
d erlaubter Raum *m*

106 alluvial soil; river valley soil
f sol *m* alluvial
e suelo *m* aluvial
i terreno *m* alluvionale
p solo *m* aluvial
d Alluvialboden *m*

107 all-weather aircraft
f avion *m* à l'épreuve des intempéries
e avión *m* para todo tiempo
i aeromobile *m* ogni-tempo
p avião *m* para qualquer condição meteorológica
d Allwetterflugzeug *n*

108 alpine *adj*
f alpin
e alpino
i alpino
p alpino
d alpin

109 alpine environment
f milieu *m* alpin
e medio *m* alpino
i ambiente *m* alpino
p ambiente *m* alpino
d alpines Milieu *n*

110 alpine pasture
f pâturage *m* alpestre
e pasto *m* alpino
i pastoreio *m* alpino
p campo *m* de pastagem alpino
d Almweide *f*; Alpweide *f*

111 alternative farming
f agriculture *f* alternative
e agricultura *f* alternativa
i agricoltura *f* alternativa
p agricultura *f* alternativa
d alternative Landwirtschaft *f*

112 altimeter
f altimètre *m*
e altímetro *m*
i altimetro *m*
p altímetro *m*
d Höhenmesser *m*

113 altimeter setting
f calage *m* de l'altimètre; calage *m* altimétrique
e ajuste *m* del altímetro
i regolazione *f* dell'altimetro
p ajuste *m* de altímetro
d Höhenmessereinstellung *f*

114 altimetry
f altimétrie *f*
e altimetría *f*
i altimetria *f*
p altimetria *f*
d Höhenmessung *f*

115 altitude
f hauteur *f*; altitude *f*
e altitud *f*
i altitudine *f*
p altitude *f*
d Höhe *f*

116 altitude error; error in height
f erreur *f* de hauteur
e error *m* de altura
i errore *m* di altezza
p erro *m* de altura
d Höhenfehler *m*

117 altitude of the sun
f hauteur *f* du soleil
e altura *f* del sol
i altezza *f* del sole
p altura *f* do sol
d Sonnenhöhe *f*

118 altitude scale
f échelle *f* des altitudes
e escala *f* de altitud
i scala *f* di altitudine
p escala *f* de altitude
d Höhenskala *f*

119 altocumulus
(a principal cloud type, usually grey or white)
f alto-cumulus *m*
e altocúmulo *m*; altocumulus *m*
i altocumulo *m*
p altocúmulo *m*; altocumulus *m*
d Altocumulus *m*

120 altocumulus castellanus
(altocumulus in which upper part presents developed projections like a tower)
f alto-cumulus *m* castellanus
e altocúmulo *m* castellanus
i altocumulo *m* castellata
p altocumulus *m* castellanus
d Altocumulus *m* castellanus

121 altocumulus floccus
(altocumulus formation characterized by small tufts or masses)
f alto-cumulus *m* floccus
e altocúmulo *m* floccus
i altocumulo *m* fiocco
p altocumulus *m* floccus
d Altocumulus *m* floccus

122 altostratus
(blush cloud formation covering the visible sky)
f alto-stratus *m*
e altoestrato *m*
i altoestrato *m*
p alto-estrato *m*
d Altostratus *m*

* **a.m. → 1521**

123 ambient air
f air *m* ambiant
e aire *m* ambiente
i aria *f* ambiente
p ar *m* ambiente
d Umgebungsluft *f*

124 ambient temperature
f température *f* ambiante
e temperatura *f* ambiente
i temperatura *f* ambiente
p temperatura *f* ambiente
d Umgebungstemperatur *f*

125 amenity planning
f aménagement *m* des espaces libres
e planificación *f* de zonas verdes y espacios libres
i progettazione *f* degli esterni
p planificação *f* de zonas verdes
d Grünplanung *f*

126 ammonia
f ammoniac *m*
e amoníaco *m*
i ammoniaca *f*
p amoníaco *m*
d Ammoniak *m*

127 amorphous ice
f glace *f* amorphe
e hielo *m* amorfo
i ghiaccio *m* amorfo
p gelo *m* amorfo
d amorphes Eis *n*

128 amount of evaporation
f quantité *f* d'évaporation
e grado *m* de evaporación
i volume *m* di evaporazione
p grau *m* de evaporação
d Verdunstungshöhe *f*

129 amount of heat
f quantité *f* de chaleur
e cantidad *f* de calor
i quantità *f* di calore
p quantidade *f* de calor
d Wärmemenge *f*

* **amount of rainfall → 258**

130 amount of water vapour
f quantité *f* de vapeur d'eau
e cantidad *f* del vapor de agua
i quantità *f* di vapore acqueo
p quantidade *f* de vapor da água
d Wasserdampfgehalt *m*

131 amplitude
f amplitude *f*
e amplitud *f*
i ampiezza *f*
p amplitude *f*
d Amplitude *f*

132 anabatic *adj*
f anabatique
e anabático
i anabatico
p anabático
d anabatisch

133 anabatic wind; up-current
(uphill wind due to surface heating)
f vent *m* anabatique; courant *m* ascendant
e viento *m* anabático; corriente *f* ascendente
i vento *m* anabatico; corrente *f* ascendente
p vento *m* anabático; corrente *f* ascendente
d Aufwind *m*

134 anaflow; airflow
f flux *m* anabatique; écoulement *m* d'air
e flujo *m* anabático; flujo *m* de aire
i flusso *m* anabatico; flusso *m* d'aria
p fluxo *m* anabático; circulação *f* de ar
d Luftstrom *m*

135 analysis
f analyse *f*
e análisis *f*
i analisi *f*
p análise *f*
d Analyse *f*

136 analysis report
f résultats *mpl* d'analyse
e boletín *m* de análisis
i risultato *m* di una analisi
p boletim *m* de análise
d Analysenergebnis *n*

137 anemobiagraph
f anémobiagraphe *m*
e anemómetro *m* a presión
i anemografo *m* aerodinamico
p anemógrafo *m* a tubo de pressão
d Anemobiagraph *m*

138 anemogram
f anémogramme *m*
e anemograma *m*
i anemogramma *m*
p anemograma *m*
d Anemogramm *n*

139 anemograph
f anémographe *m*
e anemógrafo *m*
i anemografo *m*
p anemógrafo *m*
d Anemograph *m*

140 anemography
f anémographie *f*
e anemografía *f*
i anemografia *f*
p anemografia *f*
d Anemographie *f*

141 anemology
f anémologie *f*
e anemología *f*
i anemologia *f*
p anemologia *f*
d Anemologie *f*; Wissenschaft *f* von den Luftströmungen

142 anemometer
f anémomètre *m*
e anemómetro *m*
i anemometro *m*
p anemômetro *m*
d Anemometer *n*

143 anemometric scale
f échelle *f* anémométrique
e escala *f* anemométrica
i scala *f* anemometrica
p escala *f* anemométrica
d anemometrische Skala *f*

144 anemometrograph
f anémométrographe *m*
e anemometrógrafo *m*
i anemometrografo *m*
p anemometrógrafo *m*
d Anemometrograph *m*; Windschreiber *m*

145 anemometry; wind measurement
f anémométrie *f*; mesure *f* du vent
e anemometría *f*; medición *f* del viento
i anemometria *f*; misura *f* del vento
p anemometria *f*; medição *f* do vento
d Anemometrie *f*; Windmessung *f*

146 anemoscope; wind vane
f anémoscope *m*
e anemoscopio *m*; veleta *f*
i anemoscopio *m*
p anemoscópio *m*; catavento *m*
d Anemoskop *n*

147 aneroid
f anéroïde *m*
e aneroide *m*
i aneroide *m*
p aneróide *m*
d Aneroid *n*

148 aneroid barometer
f baromètre *m* anéroïde
e barómetro *m* aneroide
i barometro *m* aneroide
p barômetro *m* aneróide
d Aneroidbarometer *n*

149 angle of downwash
f angle *m* de déflection du filament d'air
e ángulo *m* de deflexión
i angolo *m* di influsso
p ângulo *m* de perturbação abaixo
d Abwindwinkel *m*

150 Angström
(unit of length (one hundred millionth of a centimeter) to express wavelengths of light)
f Angström *m*
e Angström *m*
i Angström *m*
p Angström *m*
d Angström *n*

151 annual *adj*
f annuel
e anual
i annuale
p anual
d jährlich

152 annual layer
f couche *f* annuelle
e estrato *m* anual
i strato *m* annuale
p camada *f* anual
d Jahresschicht *f*

* **annual precipitation → 153**

153 annual rainfall; annual precipitation
f précipitation *f* annuelle
e precipitación *f* anual
i precipitazione *f* annua
p precipitação *f* anual
d Jahresniederschlag *m*

154 annual temperature
f température *f* annuelle
e temperatura *f* anual
i temperatura *f* annua
p temperatura *f* anual
d Jahrestemperatur *f*

155 annual weather
f conditions *fpl* météorologiques annuelles
e climatología *f* anual
i condizioni *fpl* meteorologiche annue
p condições *fpl* meteorológicas anuais
d Jahreswitterung *f*

156 anomalous propagation of sound
f propagation *f* anormale du son
e propagación *f* anormal del sonido
i propagazione *f* anomala del suono
p propagação *f* anormal do som
d anomale Schallausbreitung *f*

157 anomalous sound
f son *m* anormal
e sonido *m* anormal
i suono *m* anomalo
p som *m* anormal
d anomaler Schall *m*

158 antarctic *adj*
f antarctique
e antártico
i antartico
p antártico
d antarktisch

159 antarctic air
f air *m* antarctique
e aire *m* antártico
i aria *f* antartica
p ar *m* antártico
d südliche Polarluft *f*

160 antarctic circle
f cercle *m* polaire antarctique
e círculo *m* polar antártico
i circolo *m* polare antartico
p círculo *m* polar antártico
d südlicher Polarkreis *m*

*** antarctic circumpolar current → 2992**

161 antarctic oasis
f oasis *f* antarctique
e oasis *m* antártico
i oasi *f* antartica
p oásis *m* antártico
d antarktische Oase *f*

162 antarctic ocean
f océan *m* antarctique
e océano *m* antártico
i oceano *m* antartico
p oceano *m* antártico
d südliches Eismeer *n*

163 antarctic region
f région *f* antarctique
e región *f* antártica
i regione *f* antartica
p região *f* antártica
d Südpolargebiet *n*

164 antenna
f antenne *f*
e antena *f*
i antenna *f*
p antena *f*
d Antenne *f*

165 anthelion
f anthélie *m*
e antihelio *m*
i antelio *m*
p antélio *m*
d Gegensonne *f*

166 anticyclogenesis
f anticyclogenèse *f*
e anticiclogénesis *f*
i anticiclogenesi *f*
p origem *f* de um anticiclone
d Entwicklung *f* eines Antizyklons

*** anticyclone → 1363**

167 anticyclonic gloom
f obscurité *f* anticyclonique
e obscuridad *f* anticiclónica
i oscurità *f* anticiclonica
p escuridade *f* anticiclônica
d Antizyklondunkel *n*

168 antifreeze
f antigel *m*
e antihielo *m*
i antigelo *m*
p anticongelante *m*
d Frostschutzmittel *n*

169 anti-icer
f antigivrage *m*
e dispositivo *m* anticongelante
i dispositivo *m* antighiaccio
p degelador *m*
d Enteiser *m*

170 antipode
f antipode *m*
e antipoda *m*
i antipode *m*
p antipoda *m*
d Antipode *m*

171 antiselene
f antisélène *f*
e antiselena *f*
i antiselene *f*
p anti-selênio *m*
d Gegenmond *m*

172 antitrades; antitrade winds
(winterly winds prevailing at upper levels above the trade winds)
f contre-alizés *mpl*
e contra-alisios *mpl*
i controalisei *mpl*
p anti-alísios *mpl*
d Gegenpassate *mpl*; Gegenpassatwinde *mpl*

*** antitrade winds → 172**

173 antitriptic winds
f vents *mpl* antitriptiques
e vientos *mpl* antitrípticos
i venti *mpl* antitriptici
p ventos *mpl* antitrípticos
d antitriptische Winde *mpl*

*** antitwilight → 656**

174 anvil cloud
(cumulonimbus cloud suggesting the form an anvil)
f nuage *m* en enclume

e nube *f* en yunque
i nube *f* a incudine
p nuvem *f* em forma de bigorna
d Amboßwolke *f*

175 anvil head
f enclume *f*
e cabeza *f* de yunque
i testa *f* dell'incudine
p cabeça *f* de bigorna
d Amboß *m*

176 apex
f apex *m*
e ápice *m*
i apice *m*
p ápice *m*
d Apex *m*

177 aphelion
f aphélie *f*
e afelio *m*
i afelio *m*
p afélio *m*
d Aphelium *n*

178 apical *adj*
f apical
e apical
i apicale
p apical
d apikal

179 apogee
f apogée *m*
e apogeo *m*
i apogeo *m*
p apogeu *m*
d Erdferne *f*

180 apparent *adj*
f apparent
e aparente
i apparente
p aparente
d scheinbar

181 apparent altitude; virtual height
f hauteur *f* apparente
e altura *f* aparente
i altezza *f* apparente
p altura *f* aparente
d scheinbare Höhe *f*

182 apparent dip of the horizon
f dépression *f* d'horizon apparente
e depresión *f* aparente del horizonte
i depressione *f* apparente dell'orizzonte
p depressão *f* aparente do horizonte
d scheinbare Kimmtiefe *f*

183 apparent direction of the smoke
f direction *f* apparente de la fumée
e dirección *f* aparente del humo
i direzione *f* apparente del fumo
p direção *f* aparente da fumaça
d scheinbarer Zug *m* des Rauches

184 apparent diurnal direction of motion
f mouvement *m* diurne apparent
e movimiento *m* diurno aparente
i direzione *f* del movimento diurno apparente
p movimento *m* diurno aparente
d scheinbare tägliche Bewegungsrichtung *f*

185 apparent noon
f midi *m* vrai
e mediodía *m* verdadero
i mezzogiorno *m* vero
p meio-dia *m* verdadeiro
d scheinbarer Mittag *m*

186 apparent sun; true sun
f soleil *m* vrai
e sol *m* verdadero
i sole *m* vero
p sol *m* verdadeiro
d wahre Sonne *f*

187 apparent wind
f vent *m* apparent
e viento *m* aparente
i vento *m* apparente
p vento *m* aparente
d scheinbarer Wind *m*

188 Appleton layer
f couche *f* d'Appleton
e zona *f* Appleton
i strato *m* Appleton
p camada *f* de Appleton
d Appletonschicht *f*

189 approach of the storm
f approche *f* de la tourmente
e aproximación *f* de la borrasca
i avvicinamento *m* della burrasca
p aproximação *f* da tempestade
d Annähern *n* des Sturmes

190 April
f avril *m*
e abril *m*
i aprile *m*
p abril *m*
d April *m*

191 aquatic *adj*
f aquatique
e acuático
i acquatico
p aquático
d aquatisch

192 aquatic resources
f ressources *fpl* aquatiques
e recursos *mpl* acuáticos
i risorse *fpl* idriche
p recursos *mpl* hídricos
d Wasservorräte *mpl*

193 aqueous *adj*
f aqueux
e acuoso
i acqueo
p aquoso
d wasserhaltig

* **aqueous vapour → 2954**

194 aquifuge
f couche *f* imperméable
e acuífugo *m*
i livello *m* impermeabile
p aqüífuga *f*
d Grundwasserstauer *m*

195 aquosity
f aquosité *f*
e acuosidad *f*
i acquosità *f*
p aquosidade *f*
d Wässerigkeit *f*

196 arable farming
f culture *f* labourée
e cultivo *m* del campo
i coltura *f* arativa
p agricultura *f* de semeadura
d Ackerbau *m*

197 arable field relation
f proportion *f* de terres labourées
e relación *f* de tierras de cultivo
i proporzione *f* di seminativo
p proporção *f* de terras de cultivo
d Ackerflächenverhältnis *n*

198 arable meadow
f prairie *f* temporaire fauchée
e prado *m* temporario de corte
i prato *m* temporaneo da sfalcio
p prado *m* temporário de corte
d Ackerwiese *f*

199 arable pasture
f prairie *f* temporaire pâturée
e prado *m* temporario de pastar
i pascolo *m* temporaneo
p prado *m* temporário de pasto
d Ackerweide *f*

200 arboriculture
f arboriculture *f*
e arboricultura *f*
i arboricoltura *f*
p arboricultura *f*
d Gehölzzucht *f*

201 arc; bow
f arc *m*
e arco *m*
i arco *m*
p arco *m*
d Bogen *m*

202 arc of meridian
f arc *m* méridien
e arco *m* del meridiano
i arco *m* di meridiano
p arco *m* meridiano
d Meridianbogen *m*

203 arctic *adj*
f arctique
e ártico
i artico
p ártico
d arktisch

204 arctic air
f air *m* arctique
e aire *m* ártico
i aria *f* artica
p ar *m* ártico
d nördliche Polarluft *f*

205 arctic circle; north polar circle
f cercle *m* polaire
e círculo *m* polar
i circolo *m* polare
p círculo *m* polar
d nördlicher Polarkreis *m*

206 arctic front
f front *m* arctique
e frente *m* ártico
i fronte *m* artico
p frente *f* polar ártica
d Kaltluftfront *f*

207 arctic ocean
f océan *m* arctique
e océano *m* ártico

i oceano *m* artico
p oceano *m* ártico
d nördliches Eismeer *n*

208 arctic region
f région *f* arctique
e región *f* ártica
i regione *f* artica
p região *f* ártica
d Nordpolargebiet *n*

209 arctic sea smoke
f fumée *f* de mer arctique
e niebla *f* densa de los mares árticos
i fumo *m* di mare artico
p fumaça *f* do mar ártico
d arktischer Seenebel *m*

210 arc with ray structure
f arc *m* rayé; arc *m* avec structure de rayons
e arco *m* con estructura en rayos
i arco *m* con struttura a raggi
p arco *m* com estrutura em raios
d Bogen *m* mit Strahlenstruktur

211 area
f zone *f*
e área *f*
i area *f*
p área *f*
d Fläche *f*

212 argon
(one of the rare gases, atomic number 18; symbol Ar)
f argon *m*
e argón *m*
i argo *m*
p argônio *m*
d Argon *n*

213 arid *adj*
f aride
e árido
i arido
p árido
d dürr; trocken

214 arid climate
f climat *m* aride
e clima *m* árido
i clima *m* arido
p clima *m* árido
d trockenes Klima *n*

* **aridity → 855**

215 arid zone; water deficiency area
f zone *f* aride
e zona *f* árida
i zona *f* aride
p zona *f* árida
d Trockengebiet *n*

216 artificial cloud
f nuage *m* artificiel
e nube *f* artificial
i nube *f* artificiale
p nuvem *f* artificial
d künstliche Wolke *f*

217 artificial radioactivity
f radioactivité *f* artificielle
e radiactividad *f* artificial
i radioattività *f* artificiale
p radioatividade *f* artificial
d künstliche Radioaktivität *f*

218 artificial rain
f pluie *f* artificielle
e lluvia *f* artificial; pluvificación *f*
i pioggia *f* artificiale
p chuva *f* artificial
d künstlicher Regen *m*

219 artificial satellite; satellite
f satellite *m* artificiel
e satélite *m* artificial
i satellite *m* artificiale
p satélite *m* artificial
d Satellit *m*; künstlicher Trabant *m*

220 Asiatic grippe
f grippe *f* asiatique
e gripe *f* asiática
i grippe *f* asiatica
p gripe *f* asiática
d asiatische Grippe *f*

221 assimilation
f assimilation *f*
e asimilación *f*
i assimilazione *f*
p assimilação *f*
d Assimilation *f*

222 assimilation of atmospheric nitrogen
f assimilation *f* de l'azote atmosphérique
e asimilación *f* del nitrógeno atmosférico
i assimilazione *f* dell'azoto atmosferico
p assimilação *f* do nitrogênio atmosférico
d Luftstickstoffbindung *f*

223 associated air mass
f masse *f* d'air associée
e masa *f* de aire asociada
i massa *f* d'aria spostata
p massa *f* de ar associada
d mitbewegte Luftmasse *f*

224 astral *adj*
f astral
e astral
i astrale
p astral
d sternartig

225 astronomical twilight
f crépuscule *m* astronomique
e crepúsculo *m* astronómico
i crepuscolo *m* astronomico
p crepúsculo *m* astronômico
d astronomische Dämmerung *f*

226 astronomy
f astronomie *f*
e astronomía *f*
i astronomia *f*
p astronomia *f*
d Astronomie *f*

227 astrophobia
f astrophobie *f*
e astrofobia *f*
i astrofobia *f*
p astrofobia *f*
d Astrophobie *f*

228 astrophysics
f astrophysique *f*
e astrofísica *f*
i astrofisica *f*
p astrofísica *f*
d Astrophysik *f*

229 asymmetry
f asymétrie *f*
e asimetría *f*
i asimmetria *f*
p assimetria *f*
d Asymmetrie *f*

230 Atlantic Ocean
f Océan *m* atlantique
e Océano *m* atlántico
i Oceano *m* atlantico
p Oceano *m* atlântico
d atlantischer Ozean *m*

* **atmidometer → 988**

231 atmosphere
f atmosphère *f*
e atmósfera *f*
i atmosfera *f*
p atmosfera *f*
d Atmosphäre *f*

232 atmosphere of the earth; earth's atmosphere
f atmosphère *f* terrestre
e atmósfera *f* de la tierra
i atmosfera *f* della terra; atmosfera *f* terrestre
p atmosfera *f* da terra
d Erdatmosphäre *f*

233 atmospheric *adj*
f atmosphérique
e atmosférico
i atmosferico
p atmosférico
d atmosphärisch

234 atmospheric absorption
f absorption *f* atmosphérique
e absorción *f* atmosférica
i assorbimento *m* atmosferico
p absorção *f* atmosférica
d atmosphärische Absorption *f*

235 atmospheric acoustics
f acoustique *f* atmosphérique
e acústica *f* atmosférica
i acustica *f* atmosferica
p acústica *f* atmosférica
d Atmosphärenakustik *f*

236 atmospheric agents
f agents *mpl* atmosphériques
e agentes *mpl* atmosféricos
i agenti *mpl* atmosferici
p agentes *mpl* atmosféricos
d Atmosphärilien *npl*

237 atmospheric air
f air *m* atmosphérique
e aire *m* atmosférico
i aria *f* atmosferica
p ar *m* atmosférico
d atmosphärische Luft *f*

238 atmospheric altitude pressure
f pression *f* atmosphérique d'altitudes
e presión *f* atmosférica de las alturas
i pressione *f* atmosferica delle altitudini
p pressão *f* atmosférica das alturas
d Höhenlagenluftdruck *m*

239 atmospheric circulation
f circulation *f* atmosphérique
e circulación *f* atmosférica
i circolazione *f* atmosferica
p circulação *f* atmosférica
d atmosphärische Zirkulation *f*

240 atmospheric condensation
f condensation *f* atmosphérique
e condensación *f* atmosférica
i condensazione *f* atmosferica
p condensação *f* atmosférica
d Luftkondensation *f*

241 atmospheric conditions
f conditions *fpl* atmosphériques
e condiciones *fpl* atmosféricas
i condizioni *fpl* atmospheriche
p condições *fpl* atmosféricas
d Luftzustand *m*

242 atmospheric discharge
f décharge *f* atmosphérique
e descarga *f* atmosférica
i scarica *f* atmosferica
p descarga *f* atmosférica
d Luftentladung *f*

243 atmospheric disturbances
f perturbations *fpl* atmosphériques
e perturbaciones *fpl* atmosféricas
i disturbi *mpl* atmosferici
p perturbações *fpl* atmosféricas
d atmosphärische Störungen *fpl*

244 atmospheric duct
f canal *m* atmosphérique
e conducto *m* atmosférico
i dotto *m* atmosferico
p canal *m* atmosférico
d atmosphärische Luftführung *f*

245 atmospheric electricity
f électricité *f* atmosphérique
e electricidad *f* atmosférica
i elettricità *f* atmosferica
p eletricidade *f* atmosférica
d Luftelektrizität *f*

246 atmospheric equilibrium
f équilibre *m* atmosphérique
e equilibrio *m* atmosférico
i equilibrio *m* atmosferico
p equilíbrio *m* atmosférico
d Luftausgleich *m*

247 atmospheric humidity control
f réglage *m* de l'humidité atmosphérique
e regulación *f* de la humedad del aire
i regolazione *f* dell'umidità d'aria
p regulação *f* da umidade atmosférica
d Luftfeuchtigkeitsregelung *f*

248 atmospheric ice
f glace *f* atmosphérique
e hielo *m* atmosférico
i ghiaccio *m* atmosferico
p gelo *m* atmosférico
d atmosphärisches Eis *n*

249 atmospheric influence
f influence *f* atmosphérique
e influencia *f* atmosférica
i influenza *f* atmosferica
p influência *f* atmosférica
d Witterungseinfluß *m*

250 atmospheric instability
f instabilité *f* atmosphérique
e inestabilidad *f* atmosférica
i instabilità *f* atmosferica
p instabilidade *f* atmosférica
d atmosphärische Instabilität *f*; Witterungsunbeständigkeit *f*

251 atmospheric inversion
f inversion *f* atmosphérique
e inversión *f* atmosférica
i inversione *f* atmosferica
p inversão *f* atmosférica
d atmosphärische Inversion *f*

252 atmospheric layer
f couche *f* atmosphérique
e estrato *m* atmosférico
i strato *m* atmosferico
p camada *f* atmosférica; região *f* atmosférica
d atmosphärische Schicht *f*

253 atmospheric moisture; air humidity; air moisture
f humidité *f* atmosphérique; humidité *f* de l'air
e humedad *f* atmosférica; humedad *f* del aire
i umidità *f* atmosferica; umidità *f* dell'aria
p umidade *f* atmosférica; umidade *f* do ar
d Luftfeuchtigkeit *f*

254 atmospheric nitrogen
f azote *m* atmosphérique
e nitrógeno *m* atmosférico
i azoto *m* atmosferico
p nitrogênio *m* atmosférico
d Luftstickstoff *m*

255 atmospheric oxygen
f oxygène *m* atmosphérique
e oxígeno *m* atmosférico
i ossigeno *m* atmosferico
p oxigênio *m* atmosférico
d atmosphärischer Sauerstoff *m*

256 atmospheric ozone
f ozone *m* atmosphérique
e ozono *m* atmosférico
i ozono *m* atmosferico
p ozônio *m* atmosférico
d atmosphärisches Ozon *n*

257 atmospheric ozone content
f quantité *f* d'ozone atmosphérique
e contenido *m* de ozono atmosférico
i contenuto *m* d'ozono atmosferico
p conteúdo *m* de ozônio atmosférico
d Ozongehalt *m*

*** atmospheric pollution → 91**

258 atmospheric precipitation; precipitation; amount of rainfall
f précipitation *f* atmosphérique
e precipitación *f* atmosférica
i precipitazione *f* atmosferica
p precipitação *f* atmosférica
d atmosphärischer Niederschlag *m*

259 atmospheric pressure
f pression *f* atmosphérique
e presión *f* atmosférica
i pressione *f* atmosferica
p pressão *f* atmosférica
d Luftdruck *m*

260 atmospheric pressure disease
f maladie *f* de pression atmosphérique
e enfermedad *f* de presión atmosférica
i malattia *f* di pressione atmosferica
p enfermidade *f* de pressão atmosférica
d Luftdruckerkrankung *f*

261 atmospheric radiation
f rayonnement *m* atmosphérique
e radiación *f* atmosférica
i radiazione *f* atmosferica
p radiação *f* atmosférica
d atmosphärische Strahlung *f*

262 atmospheric refraction
f réfraction *f* atmosphérique
e refracción *f* atmosférica
i rifrazione *f* atmosferica
p refração *f* atmosférica
d atmosphärische Strahlenbrechung *f*

263 atmospherics
f parasites *mpl* atmosphériques
e interferencias *fpl* atmosféricas
i parassiti *mpl* atmosferici
p ruídos *mpl* parasitas
d atmosphärische Funkstörungen *fpl*

264 atmospheric stability
f stabilité *f* atmosphérique
e estabilidad *f* atmosférica
i stabilità *f* atmosferica
p estabilidade *f* atmosférica
d atmosphärische Stabilität *f*

265 atmospheric topping
f distillation *f* atmosphérique
e destilación *f* atmosférica
i distillazione *f* atmosferica
p destilação *f* atmosférica
d atmosphärische Destillation *f*

266 at night
f pendant la nuit
e de noche
i nottetempo
p à noite
d bei Nacht

267 at nightfall
f à la tombée de la nuit
e al anochecer
i al cadere della notte
p ao anoitecer
d bei Eintritt der Dunkelheit

268 atomic desert
f désert *m* atomique
e desierto *m* atómico
i deserto *m* atomico
p deserto *m* atômico
d Atomwüste *f*

269 atomic oxygen
f oxygène *m* atomique
e oxígeno *m* atómico
i ossigeno *m* atomico
p oxigênio *m* atômico
d atomarer Sauerstoff *m*

270 attached thermometer
f thermomètre *m* attaché
e termómetro *m* adjunto
i termometro *m* unito al barometro
p termômetro *m* adjunto
d attachiertes Thermometer *n*

271 audibility
f audibilité *f*
e audibilidad *f*
i audibilità *f*
p audibilidade *f*
d Hörbarkeit *f*

272 August
f août *m*
e agosto *m*

i agosto *m*
p agôsto *m*
d August *m*

273 aurora australis
f aurore *f* australe
e aurora *f* austral
i aurora *f* australe
p aurora *f* austral
d Südlicht *n*

274 aurora borealis
f aurore *f* boréale
e aurora *f* boreal
i aurora *f* boreale
p aurora *f* boreal
d Nordlicht *n*

275 auroral cap
f calotte *f* aurorale
e calota *f* auroral
i calotta *f* aurorale
p calota *f* auroral
d Polarlichtkalotte *f*

276 auroral form
f forme *f* aurorale
e forma *f* auroral
i forma *f* delle aurore
p forma *f* auroral
d Polarlichtform *f*

277 auroral frequency
f fréquence *f* aurorale
e frecuencia *f* auroral
i frequenza *f* delle aurore
p frequência *f* auroral
d Polarlichthäufigkeit *f*

278 auroral region
f région *f* aurorale
e región *f* auroral
i regione *f* aurorale
p região *f* auroral
d Polarlichtregion *f*

279 auroral type
f type *m* d'aurore
e tipo *m* de aurora
i tipo *m* dell'aurora
p tipo *m* de aurora
d Art *f* des Polarlichtes

280 auroral zone
f zone *f* aurorale
e zona *f* auroral
i zona *f* aurorale
p zona *f* auroral
d Polarlichtzone *f*

281 aurora polaris
f aurore *f* polaire
e aurora *f* polar
i aurora *f* polare
p aurora *f* polar
d Polarlicht *n*

282 austral *adj*
f austral
e austral
i australe
p austral
d südlich

283 australite
f australite *f*
e australita *f*
i australite *f*
p australita *f*
d Australit *m*

284 autoconvective lapse rate
f gradient *m* thermique autoconvectif
e gradiente *m* vertical autoconvectivo
i gradiente *m* verticale autoconvettivo
p gradiente *m* vertical autoconvectivo
d autokonvektiver Temperaturgradient *m*

285 autumn
f automne *m*
e otoño *m*
i autunno *m*
p outono *m*
d Herbst *m*

286 autumnal equinox
f équinoxe *m* d'automne
e equinoccio *m* otoñal
i equinozio *m* d'autunno
p equinócio *m* outonal
d Herbstäquinoktium *n*

287 autumnal fever
f fièvre *f* automnale
e fiebre *f* otoñal
i febbre *f* autunnale
p febre *f* outonal
d Herbstfieber *n*

288 autumn catch crop
f culture *f* dérobée d'automne
e cultivo *m* intermedio de otoño
i coltura *f* intercalare autunnale
p cultivo *m* intermediário de outono
d Herbstzwischenfrucht *f*

289 autumn colour
f couleur *f* d'automne
e color *m* otoñal

i colore *m* d'autunno
p cor *f* outonal
d Herbstfärbung *f*

290 autumn-fruiting raspberry
f framboisier *m* remontant
e frambuesa *f* con fructificación de otoño
i lampone *m* a maturazione autunnale
p framboesa *f* de frutificação outonal
d Herbsthimbeere *f*

291 auxiliary barometer
f baromètre *m* auxiliaire
e barómetro *m* auxiliar
i barometro *m* aussiliario
p barômetro *m* auxiliar
d Hilfsbarometer *n*

292 avalanche
f avalanche *f*
e alud *m*
i valanga *f*
p avalancha *f*
d Lawine *f*

293 avalanche body
f corps *m* d'avalanche
e cuerpo *m* de alud
i corpo *m* di valanga
p corpo *m* de avalancha
d Lawinenkörper *m*

294 avalanche channel
f couloir *m* d'avalanche
e canal *m* de avalancha
i canale *m* di valanga
p canal *m* de avalancha
d Lawinengleitbahn *f*

295 avalanche defense; control of avalanches
f protection *f* contre les avalanches; lutte *f* contre les avalanches
e protección *f* contra aludes
i protezione *f* contro le valanghe; difesa *f* contro le valanghe
p proteção *f* contra as avalanchas
d Kampf *m* gegen die Lawinengefahr; Lawinenschutz *m*

296 avalanche deposit
f dépôts *mpl* d'avalanche
e depósitos *mpl* de la avalancha
i depositi *mpl* della valanga
p depósitos *mpl* de avalancha
d Lawinenablagerungen *fpl*

297 avalanche front
f front *m* d'avalanche
e frente *m* de alud
i fronte *m* di valanga
p frente *f* de avalancha
d Lawinenfront *f*

298 avalanche snow
f neige *f* d'avalanche
e nieve *f* de la avalancha
i neve *f* della valanga
p neve *f* de avalancha
d Lawinenschnee *m*

299 avalanche study
f étude *f* des avalanches
e estudio *m* de las avalanchas
i studio *m* di valanghe
p estudo *m* das avalanchas
d Lawinenkunde *f*

300 avalanche train
f traîne *f* d'avalanche
e tren *m* de alud
i treno *m* di valanga
p trem *m* de avalancha
d Lawinenschleppe *f*

301 avalanche transit zone
f zone *f* de transit d'une avalanche
e zona *f* de tránsito de una avalancha
i zona *f* di transito della valanga
p zona *f* de trânsito de uma avalancha
d Sturzbahn *f* einer Lawine

*** average → 1763**

302 average temperature
f température *f* moyenne
e temperatura *f* media
i temperatura *f* media
p temperatura *f* média
d mittlere Temperatur *f*

*** average value → 1770**

303 axis of depression
f axe *m* de dépression
e eje *m* de depresión
i asse *m* di depressione
p eixo *m* de depressão
d Depressionsachse *f*

304 azimuth
f azimut *m*
e azimut *m*
i azimut *m*
p azimute *m*
d Azimut *m*

305 azimuth error
f erreur *f* d'azimut
e error *m* del azimut
i errore *m* dell'azimut
p erro *m* do azimute
d Azimutfehler *m*

B

* **backing** → 2887

306 backing of the wind
f changement *m* du vent contre le soleil
e cambio *m* de viento contra el sol
i cambiamento *m* della direzione del vento in senso antiorario
p mudança *f* de vento contra o sol
d Krimpen *n* des Windes

307 backlash
f jeu *m*
e zafada *f*
i gioco *m*
p jogo *m*
d Spielraum *m*

308 badly coloured
f mal coloré
e descolorido
i scolorito
p mal corado
d fehlfarbig

309 badoorie
(cold dry wind in the north of Australia)
f badoorie *m*
e badoorie *m*
i badoorie *m*
p badoorie *m*
d Badoorie *m*

310 bad weather aeroplane
f avion *m* de mauvais temps
e avión *m* para mal tiempo
i aeroplano *m* per tempo cattivo
p avião *m* para mau tempo
d Schlechtwetterflugzeug *n*

* **bad weather, on account of ~** → **1968**

311 baffle plate
f plaque *f* de contrevent
e placa *f* deflectora
i placca *f* di controvento
p placa *f* defletora
d Windzacken *m*

312 baffling wind
f vent *m* variable
e viento *m* rolante
i vento *m* variabile
p vento *m* variável
d umspringender Wind *m*

313 balance barometer
f baromètre *m* à balance
e barómetro *m* de balanza
i barometro *m* a bilancia
p barômetro *m* de balança
d Wagebarometer *n*

314 ball-lightning
f éclair *m* en boule
e relámpago *m* en bola
i lampo *m* sferico
p raio *m* esférico
d Kugelblitz *m*

315 balloon
f ballon *m*
e globo *m* aerostático
i pallone *m*
p balão *m*
d Ballon *m*

316 balloon-sonde; sounding balloon
f ballon-sonde *m*
e globo *m* de sondeo
i pallone *m* sonda
p balão-sonda *m*
d Registrierballon *m*

317 Baltic *adj*
f baltique
e báltico
i baltico
p báltico
d baltisch

318 band
f bande *f*
e banda *f*
i banda *f*
p banda *f*
d Bande *f*

319 band system
f système *m* de bandes
e sistema *m* de bandas
i sistema *m* di bande
p sistema *m* de bandas
d Bandensystem *n*

320 band with ray structure
f bande *f* rayée; bande *f* avec structure de rayons
e banda *f* con estructura en rayos

i banda *f* con struttura a raggi
p banda *f* com estrutura em raios
d Bande *f* mit Strahlenstruktur

321 bank
f berge *f*
e banco *m*
i banco *m* sommerso
p banco *m*
d Felsbank *f*

322 banner cloud
(a cloud extending downwind from a mountain peak in a flag-like form)
f nuage *m* en forme de drapeau
e nube *f* en forma de bandera; nube *f* bandera
i nube *f* a bandiera
p nuvem *f* en forma de bandeira
d Wolkenfahn *f*

323 bar
(unit of pressure: 1 bar = 10^5 N/m^2)
f bar *m*
e bar *m*
i bar *m*
p bar *m*
d Bar *n*

324 barat
(violent squall from the northwest blowing on Celebes)
f barat *m*
e barat *m*
i barat *m*
p barat *m*
d Barat *m*

325 barb
f barbule *f*
e barbilla *f*
i barba *f*
p barba *f*
d Bart *m*

326 barogram
f barogramme *m*
e barograma *m*
i barogramma *m*
p barograma *m*
d Barogramm *n*

327 barograph; recording barometer
f barographe *m*; baromètre *m* enregistreur
e barógrafo *m*; barómetro *m* registrador
i barografo *m*; barometro *m* registratore
p barógrafo *m*; barômetro *m* registrador
d Barograph *m*; registrierendes Barometer *n*

328 barometer
f baromètre *m*
e barómetro *m*
i barometro *m*
p barômetro *m*
d Barometer *n*; Luftdruckmesser *m*

329 barometer is falling
f le baromètre tombe
e el barómetro baja
i il barometro abbassa
p o barômetro baixa
d das Barometer fällt

330 barometer is rising
f le baromètre monte
e el barómetro sube
i il barometro sale
p o barômetro sobe
d das Barometer steigt

331 barometer is steady
f le baromètre ne bouge pas
e el barómetro está fijo
i il barometro è fisso
p o barômetro está estável
d das Barometer steht

332 barometer reading
f lecture *f* de baromètre
e lectura *f* del barómetro
i lettura *f* del barometro
p leitura *f* de barômetro
d Barometerstand *m*

333 barometer scale
f échelle *f* de baromètre
e escala *f* de barómetro
i scala *f* di barometro
p escala *f* de barômetro
d Barometerskala *f*

334 barometer variation
f variation *f* barométrique
e variación *f* barométrica
i variazione *f* barometrica
p variação *f* barométrica
d barometrische Schwankung *f*

335 barometer with corrected scale
f baromètre *m* à échelle compensée
e barómetro *m* de escala compensada
i barometro *m* a escala ridotta
p barômetro *m* de escala compensada
d Luftdruckmesser *m* mit verkleinerter Skala

336 barometric *adj*; **barometrical** *adj*
f barométrique
e barométrico

i barometrico
p barométrico
d barometrisch

* **barometrical** *adj* → **336**

337 barometric altimeter; pressure altimeter
f altimètre *m* barométrique
e altímetro *m* barométrico
i altimetro *m* a pressione
p altímetro *m* barométrico
d Druckhöhenmesser *m*

338 barometric gradient; pressure gradient
f gradient *m* de pression
e gradiente *m* bárico
i gradiente *m* di pressione
p gradiente *m* barométrico
d barometrische Luftdrucksteigung *f*; Luftdruckgradient *m*

339 barometric height formula
f formule *f* barométrique; formule *f* altimétrique
e fórmula *f* barométrica; fórmula *f* de altimetría
i formola *f* barometrica; formola *f* altimetrica
p fórmula *f* barométrica
d Barometerformel *f*; barometrische Höhenformel *f*

340 barometric observations
f observations *fpl* barométriques
e observaciones *fpl* barométricas
i osservazioni *fpl* barometriche
p observações *fpl* barométricas
d barometrische Beobachtungen *fpl*

341 barometric pressure
f pression *f* barométrique
e presión *f* barométrica
i pressione *f* barometrica
p pressão *f* barométrica
d Barometerdruck *m*

342 barometric pressure control
f correcteur *m* barométrique
e regulación *f* barométrica de presión
i controllo *m* di pressione barometrica
p controle *m* de pressão de ar
d Luftdruckregelung *f*

343 barometric pressure sensor
f capteur *m* de pression barométrique
e captador *m* de presión barométrica
i sensore *m* di pressione barometrica
p sensor *m* de pressão barométrica
d Luftdrucksensor *m*

344 barometric records
f enregistrements *mpl* barométriques
e registraciones *fpl* barométricas
i registrazioni *fpl* barometriche
p registros *mpl* barométricos
d barographische Aufzeichnungen *fpl*

345 barometric switch
f commutateur *m* barométrique
e interruptor *m* barométrico
i interruttore *m* barometrico
p interruptor *m* barométrico
d Barometerrelais *n*

346 barometric tendency
f tendance *f* barométrique
e tendencia *f* barométrica
i tendenza *f* barometrica
p tendência *f* barométrica
d Barometertendenz *f*

347 barometrograph
f barométrographe *m*
e barometrógrafo *m*
i barometrografo *m*
p barometrógrafo *m*
d Barometrograph *m*

348 baroscope
f baroscope *m*
e baroscopio *m*
i baroscopio *m*
p baroscópio *m*
d Baroskop *n*

349 barothermograph; meteorograph
f barothermographe *m*; météorographe *m*
e barotermógrafo *m*; meteorógrafo *m*
i barotermografo *m*; meteorografo *m*
p barotermógrafo *m*; meteorógrafo *m*
d Meteorograph *m*

350 barothermometer
f barothermomètre *m*
e barotermómetro *m*
i barotermometro *m*
p barotermômetro *m*
d Barothermometer *n*

351 barrier ice; ice shelf
f barrière *f* des glaces
e barrera *f* de hielos
i barriera *f* di ghiacci
p barreira *f* de gelos
d Eisbarriere *f*

352 barycentre
f barycentre *m*
e baricentro *m*

i baricentro *m*
p baricentro *m*
d Baryzentrum *n*; Schwerpunkt *m*

353 barysphere
f barysphère *f*
e barisfera *f*
i centrosfera *f*
p barisfera *f*
d Barysphäre *f*

354 basin
f bassin *m*
e escudo *m*
i bacino *m*
p bacia *f*
d Bassin *n*

355 bathymetry
f bathymétrie *f*
e batimetría *f*
i batimetria *f*
p batimetria *f*
d Bathymetrie *f*; Meerestiefenmessung *f*

356 bathythermograph
f bathythermographe *m*
e batitermógrafo *m*
i batitermografo *m*
p batitermógrafo *m*
d Bathythermograph *m*

357 bay; inlet
f baie *f*; anse *f*
e bahía *f*; enseñada *f*
i baia *f*
p baía *f*; enseada *f*; angra *f*
d Bai *f*; Meerenge *f*

358 bayamo
(violent squall blowing Bight of Bayamo, on Cuba)
f bayamo *m*
e bayamo *m*
i bayamo *m*
p bayamo *m*
d Bayamo *m*

359 bay free of ice
f baie *f* libre de glaces
e bahía *f* libre de hielo
i baia *f* libera di ghiacci
p baía *f* livre de gelo
d eisfreie Bai *f*

360 bay ice
f glace *f* de baie
e hielo *m* de bahía
i ghiaccio *m* di baia
p gelo *m* de baía
d Buchteis *n*

361 beach
f plage *f*
e playa *f*
i spiaggia *f*
p praia *f*
d Strand *m*

*** beam wind → 676**

362 Beaufort notation
f notation *f* de Beaufort
e notación *f* de Beaufort
i indicazione *f* di Beaufort
p notação *f* Beaufort
d Beaufortbezeichnung *f*

363 Beaufort scale
f échelle *f* de Beaufort
e escala *f* de vientos Beaufort
i scala *f* di Beaufort
p escala *f* de Beaufort
d Beaufortskala *f*

364 bedded plants
f plantes *fpl* repiquées
e plántulas *fpl* de semillero
i piantine *fpl* trapiantate
p plantas *fpl* de viveiro
d versetzte Pflanzen *fpl*

365 bench mark
f repère *m*
e punto *m* altimétrico
i caposaldo *m* altimetrico
p ponto *m* altimétrico
d Abrißpunkt *m*

366 Benguela Current
f Courant *m* Benguela
e Corriente *f* Benguela
i Corrente *f* Benguela
p Corrente *f* Benguela
d Benguelastrom *m*

367 bentu de soli
(strong east wind blowing on the coast of Sardinia)
f bentu de soli *m*
e bentu de soli *m*
i bentu de soli *m*
p bentu de soli *m*
d Bentu de Soli *m*

368 bergy bit
f bloc *m* montueux de glace flottante
e iceberg *m* pequeño

i frammento *m* di banchisa
p icebergue *m* pequeno
d Eisstück *n*

369 bhoot
(dust whirlwind blowing in India)
f bhoot *m*
e bhoot *m*
i bhoot *m*
p bhoot *m*
d Bhoot *m*

370 bimetallic thermometer
f thermomètre *m* bimétallique
e termómetro *m* bimetálico
i termometro *m* bimetallico
p termômetro *m* bimetálico
d Bimetallthermometer *n*

371 biochore
(environmental conditions promising at life)
f biochore *f*
e biocora *f*
i biocora *f*
p biocora *f*
d Biochore *f*

372 bioclimatics
f bioclimatique *f*
e bioclimático *m*
i bioclimatico *m*
p bioclimático *m*
d Bioklimatik *f*

373 biodegradable *adj*
f biodégradable
e biodegradable
i biodegradabile
p biodegradável
d biologisch abbaubar

374 biodynamic cultivation
f agriculture *f* biodynamique
e cultivo *m* biodinámico
i coltura *f* biodinamica
p cultura *f* biodinâmica
d biologisch-dynamische Wirtschaftsweise *f*

375 biological action of radiations
f action *f* biologique des radiations
e acción *f* biológica de las radiaciones
i azione *f* biologica delle radiazioni
p ação *f* biológica das radiações
d biologische Wirkung *f* der Strahlungen

376 biological concentration
f concentration *f* biologique
e concentración *f* biológica
i concentrazione *f* biologica
p concentração *f* biológica
d biologische Konzentration *f*

377 biological concentration factor
f facteur *m* de concentration biologique
e factor *m* de concentración biológica
i fattore *m* di concentrazione biologica
p fator *m* de concentração biológica
d biologischer Konzentrationsfaktor *m*

378 biological control
f lutte *f* biologique
e control *m* biológico
i controllo *m* biologico
p controle *m* biológico
d biologische Bekämpfung *f*

379 biological effect of radiation
f effet *m* biologique du rayonnement
e efecto *m* biológico de la radiación
i effetto *m* biologico della radiazione
p efeito *m* biológico da radiação
d biologischer Strahlungseffekt *m*

380 biological half-life
f demi-vie *f* biologique
e período *m* biológico
i tempo *m* biologico di dimezzamento
p meia-vida *f* biológica
d biologische Halbwertzeit *f*

381 biological protection
f protection *f* biologique
e protección *f* biológica
i protezione *f* biologica
p proteção *f* biológica
d biologischer Schutz *m*

382 biological shield
f bouclier *m* biologique
e blindaje *m* biológico
i schermo *m* biologico
p blindagem *m* biológica
d biologische Abschirmung *f*

383 biomass
(the total dry weight of organic matter)
f biomasse *f*
e biomasa *f*
i biomassa *f*
p biomassa *f*
d Biomasse *f*

384 biosphere
(the live zone of the earth including the hydrosphere, lithosphere and atmosphere)
f biosphère *f*
e biósfera *f*

i biosfera *f*
p biosfera *f*
d Biosphäre *f*

385 biotope
f espace *m* vital; biotope *m*
e biótopo *m*
i biotopo *m*; spazio *m* vitale
p biótopo *m*
d Biotop *m*

386 bise
(cold, dry northerly wind occurring in Southern Europe)
f bise *f*
e bisa *m*
i bise *f*
p bisa *f*
d Bise *f*

387 Bishop's ring
f anneau *m* de Bishop
e anillo *m* de Bishop
i anello *m* di Bishop
p anel *m* de Bishop
d Bishopscher Ring *m*

388 bitter cold
f froid *m* rigoureux
e frío *m* rigoroso
i freddo *m* crudo
p frio *m* rigoroso
d bittere Kälte *f*

389 black body
f corps *m* noir
e cuerpo *m* negro
i corpo *m* nero
p corpo *m* opaco
d schwarzer Körper *m*

390 black-body radiation
f rayonnement *m* d'origine thermique
e radiación *f* de origen térmico
i radiazione *f* da corpo nero
p radiação *f* de origem térmica
d Wärmestrahlung *f*

391 black-bulb thermometer
f thermomètre *m* à boule noire
e termómetro *m* de bulbo negro
i termometro *m* a bulbo annerito
p termômetro *m* de reservatório negro
d Schwarzkugelthermometer *n*

392 black ice
f glace *f* noire
e hielo *m* oscuro
i ghiaccio *m* scuro
p gelo *m* escuro
d schwarzes Eis *n*

393 Black Sea
f Mer *f* Noire
e Mar *m* Negro
i Mare *m* Nero
p Mar *m* Negro
d Schwarzes Meer *n*

394 black squall
f grain *m* noir
e chubasco *m* negro
i groppo *m* scuro
p tempestade *f* negra
d schwarze Bö *f*

395 blanch *v*
f blanchir
e descolorar
i imbianchire
p descolorar
d blanchieren

396 blaze
f blaze *f*
e fuego *m*; flama *f*
i fiamma *f*
p chama *f* brilhante
d Flamme *f*

* **blizzard → 2433**

397 blood-rain
f pluie *f* de sang
e lluvia *f* de sangre
i pioggia *f* di sangue
p chuva *f* de sangue
d Blutregen *m*

* **blossoming → 1070**

398 blossom spraying
f pulvérisation *f* pendant la floraison
e pulverización *f* floral
i irrorazione *f* floreale
p pulverização *f* floral
d Blütespritzung *f*

399 blow *v*
f venter
e ventear
i ventare
p ventar
d wehen

400 blue *adj*
f bleu
e azul

i azzurro
p azul
d blau

401 blue ice
f glace *f* bleue
e hielo *m* azul
i ghiaccio *m* azzurro
p gelo *m* azul
d Blaueis *n*

402 blue of the sky
f bleu *m* du ciel
e azul *m* del cielo
i blu *m* del cielo
p azul *m* do céu
d Himmelsblaue *f*

403 blue sun
f soleil *m* bleu
e sol *m* azul
i sole *m* blu
p sol *m* azul
d blaue Sonne *f*

404 body heat; body temperature
f chaleur *f* du corps; température *f* du corps
e calor *m* corporal
i calore *m* del corpo
p calor *m* corporal
d Körperwärme *f*; Körpertemperatur *f*

* **body temperature → 404**

405 boil *v*
f bouillir
e ebullir
i bollire
p ebulir
d kochen

406 boiling; ebullition
f ébullition *f*
e ebullición *f*
i ebollizione *f*
p ebulição *f*
d Kochen *n*; Sieden *n*

407 bolometer
f bolomètre *m*
e bolómetro *m*
i bolometro *m*
p bolômetro *m*
d Bolometer *n*

408 bolometric correction
f correction *f* bolométrique
e corrección *f* bolométrica
i correzione *f* bolometrica
p correção *f* bolométrica
d bolometrische Korrektion *f*

409 Boltzmann's constant
f constante *f* de Boltzmann
e constante *f* de Boltzmann
i costante *f* di Boltzmann
p constante *f* de Boltzmann
d Boltzmannsche Konstante *f*

410 bora
(cold wind from the north, Adriatic Sea, Black Sea)
f bora *m*
e bora *f*
i bora *f*
p bora *f*
d Bora *f*

411 boreal *adj*
f boréal
e boreal
i boreale
p boreal
d nördlich

412 boreal climate
f climat *m* boréal
e clima *m* boreal
i clima *m* boreale
p clima *m* boreal
d boreales Klima *n*

413 bottom gradient
f pente *f* du sol
e gradiente *m* del suelo
i inclinazione *f* del fondo
p gradiente *m* do solo
d Sohlengefälle *n*

414 bottom ice
f glace *f* de fond
e hielo *m* del fondo
i ghiaccio *m* di fondo
p gelo *m* do fundo
d Grundeis *n*

415 boulder clay
f argile *f* glaciale à blocaux
e arcilla *f* glacial con blocos
i argilla *f* glaciale con blocchi
p argila *f* glacial com blocos
d Grundmoränen *fpl* mit Blöcken

416 boundary layer
f couche *f* limite
e capa *f* límite
i strato *m* limite
p capa *f* limite
d Grenzschicht *f*

417 bound water in snow
f eau *f* de rétention
e agua *f* retenida de la lluvia
i acqua *f* legata della pioggia
p água *f* retida pela chuva
d Haftwasser *n* im Schnee

418 Bourdon's barometer
f baromètre *m* de Bourdon
e barómetro *m* de Bourdon
i barometro *m* Bourdon
p barômetro *m* de Bourdon
d Bourdonbarometer *n*

419 bouyancy
f force *f* ascensionnelle
e boyanza *f*
i leggerezza *f*
p carga *f* de flutuação
d Auftrieb *m*

*** bow → 201**

420 brave west winds; roaring forties
f braves vents *mpl* d'ouest; bonnes brises *fpl* de l'ouest
e vientos *mpl* bravos del oeste; vientos *mpl* fuertes del oeste
i venti *mpl* bravi di ovest
p ventos *mpl* fortes de oeste
d brave Westwinde *mpl*; stürmische Westwinde *mpl*

421 Brazil Current
f Courant *m* du Brésil
e Corriente *f* del Brasil
i Corrente *f* del Brasile
p Corrente *f* do Brasil
d Brasilstrom *m*

422 breaker
f brisant *m*
e rompiente *m*
i frangente *f*
p arrebentação *f*
d Brecher *m*; Sturzwelle *f*

423 breeze
f brise *f*
e brisa *f*
i brezza *f*
p brisa *f*
d Brise *f*

424 brightness; luminosity
f brillance *f*; luminosité *f*
e luminosidad *f*
i luminosità *f*
p luminosidade *f*
d Leuchtstärke *f*

425 bright night
f nuit *f* claire
e noche *f* brillante
i notte *f* chiara
p noite *f* clara
d helle Nacht *f*

426 British Thermal Unit
f unité *f* thermique britannique
e unidad *f* térmica británica
i British Thermal Unit
p unidade *f* térmica inglesa
d britische Wärmeeinheit *f*

427 broken clouds
f couche *f* de nuages à intervalles
e nubes *fpl* fragmentadas
i nubi *fpl* a masse isolate
p nuvens *fpl* fragmentadas
d aufgebrochene Bewölkung *f*

428 brush discharge
f décharge *f* sans flamme
e descarga *f* radiante
i scarica *f* a fiocco
p descarga *f* luminosa
d Büchsenentladung *f*

429 buffet *v* **to billows**
f lutter contre la bourrasque
e luchar con borrasca
i lottare con burrasca
p lutar contra a tempestade (no mar)
d ringen mit dem Seesturm

*** bulb barometer → 516**

430 bull's eye cloud
f nuage *m* d'œil de bœuf
e nubarrón *m* del ojo de la tempestad
i nube *f* di occhio di bue
p nuvem *f* do olho de boi
d Sturmwolke *f*

*** bumpiness → 1292**

431 bumpy air
f atmosphère *f* agitée
e aire *m* rafagoso
i aria *f* instabile
p ar *m* de rajadas
d böige Luft *f*

432 buran
(northeast windstorm in winter and a dust storm in summer from south Russia and Siberia)
f buran *m*

e buran *m*
i burano *m*
p burão *m*
d Buran *m*

433 Bureau of Weather Reports
f service *m* des bulletins météorologiques
e servicio *m* de los boletines meteorológicos
i servizio *m* dei bolletini meteorologici
p serviço *m* de boletins meteorológicos
d Wetternachrichtenbüro *n*

434 buried firn
f névé *f* enterré
e nieve *f* enterrada
i neve *f* interrata
p neve *f* enterrada
d bedeckter Firn *m*

435 buried glacier
f glacier *m* enterré
e glaciar *m* enterrado
i ghiacciaio *m* interrato
p glaciar *m* enterrado
d bedeckter Gletscher *m*

436 buried ice
f glace *f* enterrée
e hielo *m* enterrado
i ghiaccio *m* interrato
p gelo *m* enterrado
d bedecktes Eis *n*

437 buried ice vein
f filon *m* de glace
e manantial *m* de hielo
i filone *m* di ghiaccio
p filão *m* de gelo
d bedeckter Eiskeil *m*

438 burster
(cold wind from southern Australia)
f burster *m*
e burster *m*
i burster *m*
p burster *m*
d Burster *m*

439 Buys-Ballot's law
f règle *f* de Buys-Ballot
e ley *f* de Buys-Ballot
i legge *f* di Buys-Ballot
p lei *f* de Buys-Ballot
d Buys-Ballotsches Gesetz *n*; barisches Windgesetz *n*

C

440 caldera glacier
f glacier *m* de cratère
e caldeira *f* glaciar
i ghiacciaio *m* di caldera
p caldeira *f* glaciar
d Kaldergletscher *m*

441 calibration
f calibrage *m*
e calibración *f*
i calibrazione *f*
p calibragem *f*
d Kalibrierung *f*

442 California Current
f Courant *m* de Californie
e Corriente *f* de California
i Corrente *f* della California
p Corrente *f* da Califórnia
d kalifornischer Strom *m*

443 calm
f calme *m*
e calma *f*
i calma *f* di vento
p calma *f*
d Windstille *f*

444 calm air; still air
f air *m* calme
e aire *m* en calma
i aria *f* calma
p ar *m* calmo
d ruhige Luft *f*

*** calm belt → 1520**

445 calms of Cancer
f calmes *mpl* de Cancer
e calmas *fpl* de Cáncer
i calme *fpl* del Cancro
p calmas *fpl* de Câncer
d Kalmengürtel *mpl* des Krebses

446 calms of Capricorn
f calmes *mpl* de Capricorne
e calmas *fpl* de Capricornio
i calme *fpl* del Capricornio
p calmas *fpl* de Capricórnio
d Kalmengürtel *mpl* des Steinbocks

447 calorie
f calorie *f*
e caloría *f*
i caloria *f*
p caloria *f*
d Kalorie *f*

448 calorific *adj*
f calorifique
e calorífico
i calorifico
p calorífico
d wärmeerzeugend

449 calorimeter
f calorimètre *m*
e calorímetro *m*
i calorimetro *m*
p calorímetro *m*
d Wärmemesser *m*

450 calorimetry
f calorimétrie *f*
e calorimetría *f*
i calorimetria *f*
p calorimetria *f*
d Kalorimetrie *f*

451 caloritropic *adj*
f caloritrope
e calorítropo
i caloritropo
p calorítropo
d wärmesuchend

452 calving
f vêlage *m*
e fragmentación *f* del hielo
i caduta *f* di ghiaccio
p fragmentação *f* do gelo
d Kalben *n*

453 Canary Current
f Courant *m* des Canaries
e Corriente *f* de las Canarias
i Corrente *f* delle Canarie
p Corrente *f* das Canárias
d Kanarienstrom *m*

454 candle
f candela *f*
e bujía *f*
i candela *f*
p candela *f*
d Kerze *f*

455 candle ice
f glace *f* columnaire
e hielo *m* en candela

i ghiaccio *m* candeliere
p gelo *m* em agulhas
d Eiskerze *f*

456 cap
f capuchon *m*
e toca *f*
i cappa *f*
p lenticular *m*
d Kappe *f*

457 capillarity
f capillarité *f*
e capilaridad *f*
i capillarità *f*
p capilaridade *f*
d Kapillarität *f*

458 capillarity correction
f correction *f* de capillarité
e corrección *f* de capilaridad
i correzione *f* della capillarità
p correção *f* da capilaridade
d Kapillarberichtigung *f*

459 capillary fringe
f frange *f* capillaire
e borde *m* capilar
i frangia *f* capillare
p franja *f* capilar
d Kapillarsaum *m*

460 capillary rise of soil moisture
f ascension *f* capillaire de l'eau dans le sol
e ascensión *f* capilar de agua del suelo
i ascesa *f* per capillarità della acqua del suolo
p ascensão *f* capilar da água do solo
d Kapillaraszension *f*

461 capillary water in snow
f eau *f* capillaire dans la neige
e agua *f* capilar en la nieve
i acqua *f* capillare in neve
p água *f* capilar na neve
d Kapillarwasser *n* im Schnee

462 capillatus
f capillatus *m*
e capillatus *m*
i capillatus *m*
p capillatus *m*
d Capillatus *m*

463 carbon dioxide
f dioxyde *m* de carbone
e dióxido *m* de carbono
i biossido *m* di carbonio
p dióxido *m* de carbono
d Kohlendioxid *n*

464 carbon dioxide application
f enrichement *m* en dioxyde de carbone
e fertilización *f* carbónica
i concimazione *f* con biossido di carbonio
p aplicação *f* de dióxido de carbono
d Kohlendioxiddüngung *f*

465 carbon dioxide assimilation
f assimilation *f* du dioxyde de carbone
e asimilación *f* de dióxido de carbono
i assimilazione *f* di biossido di carbonio
p assimilação *f* de dióxido de carbono
d Kohlendioxidassimilation *f*

466 carbon dioxide injury
f brunissement *m* interne dû au dioxyde de carbone
e pardeamiento *m* interno por dióxido de carbono
i danno *m* causato per il biossido di carbonio
p acastanhamento *m* interno devido ao dióxido de carbono
d Kohlendioxidfleischbräune *f*

467 carbon monoxide
f oxyde *m* de carbone
e óxido *m* de carbono
i ossido *m* di carbonio
p óxido *m* de carbono; monóxido *m* de carbono
d Kohlenoxid *n*; Kohlenmonoxid *n*

468 carbon/nitrogen ratio; C/N
f rapport *m* carbone/azote
e relación *f* carbono/nitrógeno
i rapporto *m* carbonio/azoto
p relação *f* carbono/azôto
d Kohlenstoff/Stickstoffverhältnis *n*

469 card graduated in degrees
f rose *f* graduée en degrés
e rosa *f* graduada en grados
i rosa *f* graduata in gradi
p rosa *f* graduada em graus
d Gradrose *f*

470 cardinal point
f point *m* cardinal
e punto *m* cardinal
i punto *m* cardinale
p ponto *m* cardinal
d Kardinalpunkt *m*

471 Caribbean
f Mer *f* des Caraïbes
e Mar *m* de las Caraibas
i Mare *m* dei Caraibi
p Mar *m* das Caraíbas
d karibisches Meer *n*

472 cascade
f cascade *f*
e cascada *f*
i cascata *f*
p cascata *f*
d Kaskade *f*

473 castellanus
(cloud formation with an appearance of turrets)
f castellanus *m*
e castellanus *m*
i castellanus *m*
p castellanus *m*
d Castellanus *m*

474 cataract
f cataracte *f*
e catarata *f*
i cateratta *f*
p catarata *f*; queda *f* d'aqua; cachoeira *f*
d Katarakt *m*

475 catastrophic avalanche
f avalanche *f* catastrophique
e avalancha *f* catastrófica
i valanga *f* catastrofica
p avalancha *f* catastrófica
d Schadenlawine *f*

*** catchment area → 829**

476 catchment of water
f captage *f* des eaux
e captación *f* de las aguas
i captazione *f* dell'acqua
p captação *f* das águas
d Fassen *n* von Wasser

477 cave glacier
f glacier *m* de grotte
e glaciar *m* de caverna
i ghiacciaio *m* di caverna
p glaciar *m* de caverna
d Höhlengletscher *m*

478 cave hoarfrost
f givre *m* des cavernes
e escarcha *f* de caverna
i brina *f* di caverna
p orvalho *m* de caverna
d Höhlenreif *m*

479 cave ice
f glace *f* des cavernes
e hielo *m* de cavernas
i ghiaccio *m* di cavernas
p gelo *m* de cavernas
d Höhleneis *n*

480 ceiling of the atmosphere; top of the atmosphere
f limite *f* supérieure de l'atmosphère
e límite *m* superior de la atmósfera
i limite *m* superiore dell'atmosfera
p limite *m* superior da atmosfera
d Atmosphärengrenze *f*

481 ceilometer
f plafonneur *m*
e ceilómetro *m*
i nefoipsometro *m*; strumento *m* per misura altezza nubi
p tetômetro *m*
d Wolkenhöhenmesser *m*

482 celestial equator
f équateur *m* céleste
e ecuador *m* astronómico
i equatore *m* celeste
p equador *m* celeste
d Himmelsäquator *m*

483 celestial globe; celestial sphere; firmament
f voûte *f* céleste; sphère *f* céleste; firmament *m*
e bóveda *f* celeste; firmamento *m*
i volta *f* celeste; sfera *f* celeste; firmamento *m*
p esfera *f* celeste; globo *m* celeste; firmamento *m*
d Himmelswölbung *f*; Himmelskugel *f*; Firmament *n*

484 celestial hemisphere
f hémisphère *m* céleste
e hemisferio *m* celeste
i emisfero *m* celeste
p hemisfério *m* celeste
d Himmelshalbkugel *f*

485 celestial meridian
f méridien *m* céleste
e meridiano *m* celeste
i meridiano *m* celeste
p meridiano *m* celeste
d Himmelsmeridian *m*

486 celestial poles
f pôles *mpl* célestes
e polos *mpl* celestes
i poli *m* celesti
p pólos *mpl* celestes
d Himmelspole *mpl*

*** celestial sphere → 483**

*** Celsius scale → 488**

487 centibar
(unit of pressure equivalent to 0.01 bar)
f centibar *m*
e centibar *m*
i centibar *m*
p centibar *m*
d Zentibar *n*

488 centigrade scale; Celsius scale
f échelle *f* centigrade; échelle *f* Celsius
e escala *f* en centígrados; escala *f* Celsius
i scala *f* centigrada; scala *f* Celsius
p escala *f* centígrada; escala *f* Celsius
d Celsiusskala *f*

489 centre of action
f centre *m* d'action
e centro *m* de acción
i centro *m* d'azione
p centro *m* de ação
d Aktionszentrum *n*

490 centre of pressure
f centre *m* de poussée
e centro *m* de presión
i centro *m* di pressione
p centro *m* de pressão
d Druckpunkt *m*

491 chain lightning
f éclair *m* sinueux
e relámpagos *mpl* en cadena
i lampi *mpl* a catena
p cadeia *f* de relâmpagos
d Kettenblitze *mpl*

492 changeable climate
f climat *m* variable
e clima *m* variable
i clima *m* variabile
p clima *m* variável
d veränderliches Klima *n*

493 change of monsoon
f renversement *m* de la mousson
e cambio *m* del monzón
i cambiamento *m* del monsone
p mudança *f* de monção
d Monsunwechsel *m*

494 change of temperature
f changement *m* de température
e cambio *m* de temperatura
i cambio *m* di temperatura
p mudança *f* de temperatura
d Temperaturwechsel *m*

495 change of tide; turn of the tide
f renverse *f* de la marée
e cambio *m* de marea
i cambramento *m* di marea
p mudança *f* da maré
d Flutwechsel *m*

496 change of weather; weather change
f changement *m* de temps
e cambio *m* del tiempo
i cambiamento *m* del tempo
p mudança *f* de tempo
d Wetterwechsel *m*

497 channel
f chenal *m*
e canal *m*
i canale *m*
p canal *m*
d Kanal *m*

498 channelled avalanche
f avalanche *f* de couloir
e avalancha *f* de coladero
i valanga *f* striata
p avalancha *f* canalizada
d Runsenlawine *f*

499 characteristic velocity
f vitesse *f* caractéristique
e velocidad *f* característica
i velocità *f* caratteristica
p velocidade *f* característica
d charakteristische Geschwindigkeit *f*

*** chemical hygrometer → 9**

500 chemiluminiscence
f chémiluminescence *f*
e quimioluminiscencia *f*
i chemiluminescenza *f*
p quimiluminescência *f*
d Chemolumineszenz *f*

501 chemistry of ice
f chimie *f* de la glace
e química *f* del hielo
i chimica *f* del ghiaccio
p química *f* do gelo
d Eischemie *f*

502 chemosphere
f chémosphère *f*
e quimosfera *f*
i chemosfera *f*
p quimosfera *f*
d Chemosphäre *f*

503 chilling injury
f dégât *m* de froid
e daño *m* de frío

i danno *m* dal freddo
p dano *m* por frio
d Kälteschaden *m*

504 chilling requirement
f besoin *m* en froid
e necesidad *f* de frío
i fabbisogno *m* in freddo
p necessidade *f* de frio
d Kältebedürfnis *n*

505 chionosphere
f chionosphère *f*
e quionosfera *f*
i chionosfera *f*
p quionosfera *f*
d Chionosphäre *f*

506 chocolatero
(cold wind from north of Mexico)
f chocolatero *m*
e chocolatero *m*
i chocolatero *m*
p chocolatero *m*
d Chocolatero *m*

507 cierzo
(dry, cold wind from north of Spain)
f cierzo *m*
e cierzo *m*
i cierzo *m*
p cierzo *m*
d Cierzo *m*

508 circle
f cercle *m*
e círculo *m*
i circolo *m*
p círculo *m*
d Kreislauf *m*

509 circulating air
f air *m* de circulation
e aire *m* de circulación
i aria *f* di circolazione
p ar *m* de circulação
d umlaufende Luft *f*

510 circumhorizontal arc
f arc *m* circumhorizontal
e arco *m* circunhorizontal
i cerchio *m* orizzontale
p arco *m* circunhorizontal
d Horizontalkreis *m*

511 circumzenithal arc
f arc *m* circumzénithal
e arco *m* circunzenital
i arco *m* circumzenitale
p arco *m* circunzenital
d Zirkumzenitalbogen *m*

512 cirriform *adj*
f en forme de cirrus
e cirriforme
i cirriforme
p cirriforme
d zirrusförmig

513 cirrocumulus
(type of cloud consisting of thin white patches composed of small granules or ripples)
f cirro-cumulus *m*
e cirrocumulus *m*
i cirrocumulo *m*
p cirrocúmulo *m*
d Zirrokumulus *m*

514 cirrostratus
(whitish cloud made of ice crystals producing halo phenomena)
f cirro-stratus *m*
e cirrostratus *m*
i cirrostrato *m*
p cirrostratus *m*
d Zirrostratus *m*

515 cirrus
(white cloud made of ice crystals)
f cirrus *m*
e cirro *m*
i cirro *m*
p cirro *m*
d Zirrus *m*

516 cistern barometer; bulb barometer
f baromètre *m* à cuvette
e barómetro *m* de cubeta
i barometro *m* a vaschetta
p barômetro *m* de cuba
d Gefäßluftdruckmesser *m*

517 civil twilight
f crépuscule *m* civil
e crepúsculo *m* civil
i crepuscolo *m* civile
p crepúsculo *m* civil
d bürgerliche Dämmerung *f*

518 clear ice; transparent ice
f glace *f* transparente
e hielo *m* transparente
i ghiaccio *m* trasparente
p gelo *m* transparente
d Klareis *n*

* **clearness → 2856**

519 clear sky
f ciel *m* serein
e cielo *m* despejado
i cielo *m* limpido
p céu *m* limpo
d heiterer Himmel *m*

520 clear weather
f temps *m* clair
e tiempo *m* claro
i tempo *m* chiaro
p tempo *m* claro
d sichtiges Wetter *n*

521 clear weather transmission
f transmission *f* de temps clair
e transmisión *f* de tiempo claro
i trasmissione *f* di tempo chiaro
p transmissão *f* de tempo claro
d Klarwettersendung *f*

522 climacophobia
f climacophobie *f*
e climacofobia *f*
i climacofobia *f*
p climacofobia *f*
d Klimakophobie *f*

523 climate
f climat *m*
e clima *m*
i clima *m*
p clima *m*
d Klima *n*

524 climate computer
f ordinateur *m* de régulation climatique
e calculador *m* climático
i calcolatore *m* dei dati climatici
p computador *m* climatológico
d Klimacomputer *m*

525 climate control
f contrôle *m* des températures
e control *m* climático
i controllo *m* ambientale
p regulação *f* ambiental
d Klimasteuerung *f*

526 climatic *adj*
f climatique
e climático
i climatico
p climático
d klimatisch

527 climatic changes
f variations *fpl* climatiques
e cambios *mpl* climáticos
i cambiamenti *mpl* di clima
p variações *fpl* climáticas
d Klimaänderungen *fpl*

528 climatic disease
f maladie *f* climatique
e enfermedad *f* climática
i malattia *f* climatica
p enfermidade *f* climática
d Klimakrankheit *f*

529 climatic effect
f action *f* climatique
e acción *f* climática
i azione *f* climatica
p efeito *m* climático
d Klimawirkung *f*

530 climatic factor
f facteur *m* climatique
e factor *m* climático
i fattore *m* climatico
p fator *m* climático
d klimatischer Faktor *m*

531 climatic snowline
f limite *f* climatique de la neige
e línea *f* climática de la nieve
i linea *f* climatica della neve
p linha *f* climática de la neve
d klimatische Schneegrenze *f*

532 climatic zone
f zone *f* climatique
e zona *f* climática
i zona *f* climatica
p zona *f* climática
d Klimazone *f*

533 climatological weather data
f données *fpl* climatiques
e valores *mpl* climáticos
i dati *mpl* climatici
p dados *mpl* climáticos
d Witterungswerte *mpl*

534 climatology
f climatologie *f*
e climatología *f*
i climatologia *f*
p climatologia *f*
d Klimatologie *f*; Klimakunde *f*

535 climatotherapy
f climatothérapie *f*
e climatoterapia *f*
i climatoterapia *f*
p climatoterapia *f*
d Klima(wechsel)therapie *f*

536 clinometer
f clinomètre *m*
e clinómetro *m*
i clinometro *m*
p clinômetro *m*
d Klinometer *n*

537 clip *v*
f tailler
e cortar
i tagliare
p podar
d schneiden

538 cloud
f nuage *m*
e nube *f*
i nube *f*
p nuvem *f*
d Wolke *f*

* **cloud amount → 550**

539 cloud and collision warning
f radar *m* météorologique et anti-collision
e radar *m* anticolisión y meteorológico
i radar *m* anticollisione e per avvistamento meteore
p radar *m* meteorológico e de anti-colisão
d Nebel- und Antikollisionsradar *m*

540 cloud and collision warning system
f système *m* d'avertissement de nuages et de collision
e sistema *m* de prevención de nubes y de colisiones
i sistema *m* di avvertimento nuvole e collisioni
p sistema *m* de prevenção contra nuvens e colisões
d Wolken- und Zusammenstoßwarnanlage *f*

541 cloud bank
f banc *m* de nuages
e banco *m* de nubes
i banco *m* di nubi
p banco *m* de nuvens
d Wolkenbank *f*

542 cloud base
f base *f* de nuages
e base *f* de nubes
i base *f* di nubi
p base *f* de nuvens
d Wolkenbasis *f*

543 cloud cap
f chapeau *m* de nuages
e nube *f* obturadora
i cappuccio *m* di nubi
p nuvem *f* capucho
d Wolkenhoch *n*

544 cloud chamber
f chambre *f* à nuage
e cámara *f* de niebla; cámara *f* de Wilson
i camera *f* a nebbia
p câmara *f* de nevoeiro; câmara *f* de nuvem eletrônica
d Nebelkammer *f*

545 cloud column
f colonne *f* de fumée
e columna *f* de humo
i colonna *f* di fumo
p coluna *f* de fumo
d Rauchsäule *f*

* **cloud cover → 550**

546 cloud detection radar
f radar *m* de détection de nuages
e radar *m* para la detección de nubes
i radar *m* per il rilievo delle nubi
p radar *m* de detecção de nuvens
d Wolkenradar *n*

547 cloud drip; fog drip
f pluie *f* de brouillard
e lluvia *f* de niebla
i pioggia *f* di nebbia
p chuva *f* de nevoeiro
d Nebelregen *m*

548 clouded sky; overcast sky; constable sky
f ciel *m* nuageux; ciel *m* couvert
e cielo *m* cubierto
i cielo *m* con cumuli abbondanti
p céu *m* encoberto
d wolkiger Himmel *m*; bedeckter Himmel *m*

549 cloud height
f hauteur *f* des nuages
e altura *f* de nubes
i altezza *f* delle nubi
p altura *f* das nuvens
d Wolkenhöhe *f*

550 cloudiness; nebulosity; cloud amount; cloud cover
f nébulosité *f*
e nubosidad *f*
i nuvolosità *f*
p nebulosidade *f*
d Bewölkung *f*; Nebelhülle *f*

551 cloud layer; stratum of clouds
f couche *f* de nuages
e capa *f* de nubes
i strato *m* di nubi
p camada *f* de nuvens
d Wolkenschicht *f*

552 cloudless *adj*
f sans nuages
e sin nubes
i senza nuvole
p sem nuvens
d wolkenlos

553 cloud mass
f masse *f* de nuages
e masa *f* de nubes
i massa *f* di nubi
p massa *f* de nuvens
d Wolkenmasse *f*

554 cloud point
f point *m* de figeage
e punto *m* de obscuridad
i punto *m* d'intorbidamento
p ponto *m* de névoa
d Trübungspunkt *m*

555 clouds at cold fronts
f nuages *mpl* de front froid
e nubes *fpl* de los frentes fríos
i nubi *fpl* di fronte freddo
p nuvens *fpl* de frentes frias
d Wolken *fpl* an Kaltfronten

556 cloud scale
f échelle *f* de nébulosité
e escala *f* de nubosidad
i scala *f* di nuvolosità
p escala *f* de nebulosidade
d Wolkenskala *f*

557 cloud searchlight
f projecteur *m* néphoscopique
e buscador *m* nefoscópico
i proiettore *m* per altezza nubi
p projetor *m* nefoscópico
d Wolkenscheinwerfer *m*

558 cloud seeding
f ensemencement *m* dans les nuages
e sembrado *m* de nubes
i semina *f* delle nubi
p poeira *f* atmosférica
d atmosphärischer Staub *m*

* **cloud streets → 1323**

559 cloudy *adj*
f nuageux
e nublado
i nuvoloso
p nublado
d bewölkt

560 cloudy weather
f temps *m* nuageux
e tiempo *m* nublado
i tempo *m* nuvoloso
p tempo *m* nublado
d bewölktes Wetter *n*

* **C/N → 468**

561 coalescence
f coalescence *f*
e coalescencia *f*
i coalescenza *f*
p coalescência *f*
d Koaleszenz *f*

562 coastal dune
f dune *f* cotière
e duna *f* costera
i duna *f* costiera
p duna *f* costeira
d Küstendüne *f*

563 coastal icing
f glaçage *m* cotier
e hielo *m* de la costa
i gelata *f* costiera
p banquisa *f* costeira
d Ufervereisung *f*

564 coding
f codage *m*
e codificación *f*
i codifica *f*
p codificação *f*
d Kodierung *f*

565 coefficient of tide
f coefficient *m* de marée
e coeficiente *m* de marea
i coefficiente *m* di marea
p coeficiente *m* de maré
d Gezeitenkoeffizient *m*

566 colatitude
f colatitude *f*
e colatitud *f*
i colatitudine *f*
p co-latitude *f*
d Breitenkomplement *n*

567 cold *adj*
f froid
e frío
i freddo
p frio
d kalt

568 cold air
f air *m* froid
e aire *m* frío
i aria *f* fredda
p ar *m* frio
d Kaltluft *f*

569 cold air drying
f séchage *m* par air froid
e desecación *f* por aire frío
i essiccazione *f* artificiale ad aria fredda
p dessecação *f* por ar frio
d Kaltlufttrocknung *f*

570 cold air mass
f masse *f* d'air froid
e masa *f* de aire frío
i massa *f* d'aria fredda
p massa *f* de ar frio
d Kaltluftmasse *f*

571 cold chain
f chaîne *f* du froid
e cadena *f* del frío
i catena *f* del freddo
p cadeia *f* de frio
d Kühlkette *f*

572 cold climate
f climat *m* froid
e clima *m* frío
i clima *m* freddo
p clima *m* frio
d kaltes Klima *n*

573 cold crop
f culture *f* de serre froide
e cultivo *m* sin calefacción
i coltura *f* senza riscaldamento
p cultura *f* sem aquecimento
d ungeheizte Kultur *f*

574 cold current
f courant *m* froid
e corriente *f* fría
i corrente *f* fredda
p corrente *f* fria
d kalter Strom *m*

575 cold environment
f milieu *m* froid
e medio *m* frío
i ambiente *m* freddo
p ambiente *m* frio
d Kaltzone *f*

576 cold firn zone
f zone *f* de névé froid
e zona *f* de nieve fría
i zona *f* di neve fredda
p zona *f* de neve fria
d kalte Firnzone *f*

577 cold frame
f châssis *m* froid
e cama *f* fría
i bancale *m* non riscaldato
p estufim *m*
d kalter Kasten *m*

578 cold front
f front *m* froid
e frente *m* frío
i fronte *m* freddo
p frente *f* fria
d Kaltfront *f*

579 cold glacier
f glacier *m* froid
e glaciar *m* frío
i ghiacciaio *m* freddo
p glaciar *m* frio
d kalter Gletscher *m*

580 cold glasshouse
f serre *f* froide
e invernadero *m* sin calefacción
i serra *f* fredda
p estufa *f* sem aquecimento
d Kalthaus *n*

581 cold occlusion
f occlusion *f* froide; occlusion *f* à caractère de front froid
e oclusión *f* fría; oclusión *f* con carácter de frente frío
i occlusione *f* fredda
p oclusão *f* fria
d Kaltokklusion *f*

582 cold ocean current
f courant *m* marin froid
e corriente *f* marina fría
i corrente *f* oceanica fredda
p corrente *f* marítima fria
d kalte Meeresströmung *f*

583 cold resistance; cold tolerance; winter hardiness
f résistance *f* au froid
e resistencia *f* al frío

i tolleranza *f* del freddo; resistenza *f* al freddo
p tolerância *f* ao frio; resistência *f* ao frio
d Kältetoleranz *f*; Kälteresistenz *f*

584 cold sensation
f sensation *f* de froid
e sensación *f* de frío
i sensazione *f* di freddo
p sensação *f* de frio
d Kältesensation *f*

585 cold storage injury
f altération *f* due au froid
e deterioro *m* debido al frío
i danno *m* di frigorifero
p dano *m* devido ao frio
d Kaltlagerschaden *m*

* **cold tolerance → 583**

586 cold wave
f vague *f* froide
e ola *f* de frío
i onda *f* fredda
p onda *f* de frio
d Kältewelle *f*

587 cold weather
f temps *m* froid
e tiempo *m* frío
i tempo *m* freddo
p tempo *m* frio
d kaltes Wetter *n*

588 colloid
f colloïde *m*
e coloide *m*
i colloide *m*
p colóide *m*
d Kolloid *n*

589 colloidal *adj*
f colloïdal
e coloidal
i colloidale
p coloidal
d kolloidal

590 coloured snow
f neige *f* colorée
e nieve *f* colorida
i neve *f* colorata
p neve *f* colorida
d gefärbter Schnee *m*

591 colour temperature
f température *f* chromatique
e temperatura *f* cromática
i temperatura *f* di colore
p temperatura *f* de acordo com a cor
d Farbtemperatur *f*

592 communication centre
f centre *m* de communication
e centro *m* de comunicación
i centro *m* di comunicazione
p centro *m* de comunicação
d Fernmeldezentrale *f*

593 compass
f boussole *f*; compas *m*
e brújula *f*
i bussola *f*; compasso *m*
p bússola *f*
d Kompaß *m*

594 compass bearing
f relèvement *m* de compas
e marcación *f* de brújula
i rilievo *m* del compasso
p marcação *f* de bússola
d Kompaßpeilung *f*

595 compass card; wind rose
f carte *f* de la rose
e rosa *f* de la brújula
i rosa *f* della bussola
p rosa *f* dos ventos
d Kompaßrose *f*

596 compass error
f erreur *f* de compas
e error *m* de brújula
i errore *m* di bussola
p erro *m* de bússola
d Kompaßfehler *m*

597 compensation
f compensation *f*
e compensación *f*
i compensazione *f*
p compensação *f*
d Kompensation *f*

598 compound tide
f onde *f* composée
e onda *f* compuesta
i onda *f* composta
p onda *f* composta
d Verbundtide *f*

599 compressibility
f compressibilité *f*
e compresibilidad *f*
i compressibilità *f*
p compressibilidade *f*
d Kompressibilität *f*

600 compressive strength of snow
f résistance *f* à la compression de la neige
e resistencia *f* a la compresión de la nieve
i resistenza *f* a compressione della neve
p resistência *f* à compressão da neve
d Druckfestigkeit *f* des Schnees

601 condensation
f condensation *f*
e condensación *f*
i condensazione *f*
p condensação *f*
d Kondensation *f*

602 condensation hygrometer
f hygromètre *m* à condensation
e higrómetro *m* de condensación
i igrometro *m* a condensazione
p higrômetro *m* de condensação
d Kondensationshygrometer *n*

603 condensation level
f niveau *m* de condensation
e altura *f* de condensación
i livello *m* di condensazione
p nível *m* de condensação
d Kondensationspunkt *m*

604 condensation nucleus
f noyau *m* de condensation
e núcleo *m* de condensación
i nucleo *m* di condensazione
p núcleo *m* de condensação
d Kondensationskern *m*

605 condensation trail; contrail
f traînée *f* de condensation
e estela *f* de condensación
i scia *f* di condensazione
p trilha *f* de condensação
d Kondensstreifen *m*

606 condense *v*
f se condenser
e condensarse
i condensarsi
p condensar-se
d kondensieren

607 conduction
f conduction *f*
e conducción *f*
i conduzione *f*
p condução *f*
d Leitung *f*

608 conductive heat
f chaleur *f* conductive
e calor *m* conductivo
i calore *m* conduttore
p calor *m* condutivo
d konduktive Wärme *f*

609 cone of escape
f cône *m* de libération
e cono *m* de escape
i cono *m* di fuga
p cone *m* de escape
d Entweichungskegel *m*

610 confined air
f air *m* confiné
e aire *m* confinado
i aria *f* confinata
p ar *m* confinado
d eingeschlossene Luft *f*

611 confluence
f confluence *f*
e confluencia *f*
i confluenza *f*
p confluência *f*
d Zusammensfluß *m*

612 confluence of glaciers
f confluence *f* de glaciers
e confluencia *f* de hielos
i confluenza *f* di ghiacci
p confluência *f* de gelos
d Zusammenfluß *m* von Gletschern

613 confluent glacier
f glacier *m* de confluence
e glaciar *m* confluente
i ghiacciaio *m* confluente
p glaciar *m* confluente
d Konfluenzgletscher *m*

614 congelation ice
f glace *f* de congélation
e hielo *m* de congelación
i ghiaccio *m* di congelazione
p gelo *m* de congelação
d Kongelationseis *n*

*** congestus → 695**

615 conservation
f conservation *f*
e conservación *f*
i conservazione *f*
p conservação *f*
d Bewahrung *f*

*** conservation of nature → 1901**

616 consolidation of ice
f renforcement *m* de la glace
e consolidación *f* del hielo
i consolidamento *m* del ghiaccio
p consolidação *f* do gelo
d Eisfestigung *f*

* **constable sky → 548**

617 constant level chart
f carte *f* à niveau fixe
e carta *f* de nivel constante
i carta *f* a quota costante
p carta *f* de nível constante
d Isohypsenkarte *f*

618 constant pressure
f pression *f* constante
e presión *f* constante
i pressione *f* costante
p pressão *f* constante
d Gleichdruck *m*

619 constant pressure chart
f carte *f* à pression constante
e mapa *m* de presión constante
i carta *f* a pressione costante
p mapa *m* de pressão constante
d Karte *f* konstanten Druckes

620 constant temperature
f température *f* constante
e temperatura *f* constante
i temperatura *f* costante
p temperatura *f* constante
d konstante Temperatur *f*

621 constant wind; steady wind
f vent *m* constant
e viento *m* constante
i vento *m* costante
p vento *m* constante
d gleichmäßiger Wind *m*; stetiger Wind *m*

622 contamination meter
f contaminamètre *m*
e aparato *m* medidor de la contaminación
i apparecchio *m* di misura della contaminazione
p medidor *m* de poluição
d Kontaminationsmesser *m*

623 continent
f continent *m*
e continente *m*
i continente *m*
p continente *m*
d Kontinent *m*

624 continental *adj*
f continental
e continental
i continentale
p continental
d kontinental

625 continental air
f air *m* continental
e aire *m* continental
i aria *f* continentale
p ar *m* continental
d Kontinentalluft *f*

626 continental climate
f climat *m* continental
e clima *m* continental
i clima *m* continentale
p clima *m* continental
d Kontinentalklima *n*

627 continental drift; drift of continents
f dérive *f* des continents
e deriva *f* continental
i deriva *f* dei continenti
p migração *f* dos continentes
d Kontinentenverschiebung *f*

628 continental glaciation
f glaciation *f* continentale
e glaciación *f* continental
i glaciazione *f* continentale
p glaciação *f* continental
d Kontinentalvereisung *f*

629 continental glacier; inland ice; plateau glacier
f inlandsis *m*; glacier *m* continental
e glaciar *m* continental
i ghiacciaio *m* continentale
p glaciar *m* continental
d Inlandeis *n*; Kontinentaleisdecke *f*

630 continental hemisphere
f hémisphère *m* continental
e hemisferio *m* continental
i emisfero *m* continentale
p hemisfério *m* continental
d Kontinentalhemisphäre *f*

631 continental slope
f talus *m* continental
e vertiente *f* continental
i scarpata *f* continentale
p vertente *f* continental
d Kontinentalabhang *m*

632 continuity equation
f équation *f* de continuité
e ecuación *f* de continuidad
i equazione *f* di continuità
p equação *f* de continuidade
d Kontinuitätsgleichung *f*

633 continuous *adj*
f continu
e continuo
i continuo
p contínuo
d kontinuierlich

634 contour line; isohypse
f isohypse *f*
e isohipsa *f*
i isoipsa *f*
p isoipsa *f*
d Isohypse *f*

*** contrail → 605**

635 control area
f région *f* de contrôle
e área *f* de control
i regione *f* di controllo
p área *f* de controle
d Flugsicherungskontrollbezirk *m*

636 controlled atmosphere storage
f entreposage *m* en atmosphère contrôlée
e conservación *f* en atmósfera controlada
i conservazione *f* in atmosfera controllata
p conservação *f* em atmosfera condicionada
d Gaslagerung *f*

*** control of avalanches → 295**

637 control system
f système *m* de lutte
e sistema *m* de control
i sistema *m* di controllo
p sistema *m* de proteção
d Bekämpfungsmethode *f*

638 convection
f convection *f*
e convección *f*
i convezione *f*
p convecção *f*
d Konvektion *f*

639 convection clouds
f nuages *mpl* de convection
e nubes *fpl* de convección
i nubi *fpl* di convezione
p nuvens *fpl* formadas por convecção
d Konvektionswolken *fpl*

640 convection current
f courant *m* de convection
e corriente *f* de convección
i corrente *f* di convezione
p corrente *f* de convecção
d Konvektionsstrom *m*

*** convective equilibrium → 33**

641 convergence
f convergence *f*
e convergencia *f*
i convergenza *f*
p convergência *f*
d Konvergenz *f*

642 cool *v*; **refrigerate** *v*
f réfrigérer
e refrigerar
i raffreddare
p refrigerar
d kühlen

643 coolant; refrigerant
f agent *m* frigorifique
e refrigerante *m*
i refrigerante *m*
p refrigerante *m*
d Kaltreagens *n*

644 cooling *(common cold)*
f refroidissement *m*
e resfriado *m*
i raffreddore *m*
p resfriado *m*
d Erkältung *f*

645 core-sample
f carotte *f* de sondage
e sonda *f* de muestra
i carota *f* di prelivo
p sonda *f* de amostragem
d Bohrkern *m*

646 coriolis force
f force *f* de coriolis
e aceleración *f* de coriolis
i forza *f* di coriolis
p força *f* de coriolis
d Corioliskraft *f*

647 corona
f couronne *f*
e corona *f*
i corona *f*
p coroa *f*
d Korona *f*

648 corpuscular radiation
f rayonnement *m* corpusculaire
e radiación *f* corpuscular
i radiazione *f* corpuscolare
p radiação *f* corpuscular
d Korpuskularstrahlung *f*

649 corrasion; mechanical erosion
f corrasion *f*
e corrasión *f*
i corrasione *f*
p corrasão *f*
d Korrasion *f*

650 correlation
f corrélation *f*
e correlación *f*
i correlazione *f*
p correlação *f*
d Korrelation *f*

651 correlation coefficient
f coefficient *m* de corrélation
e coeficiente *m* de correlación
i coefficiente *m* di correlazione
p coeficiente *m* de correlação
d Korrelationskoeffizient *m*

652 corrosion
f corrosion *f*
e corrosión *f*
i corrosione *f*
p corrosão *f*
d Korrosion *f*

653 cosmic dust
f poussière *f* cosmique
e polvo *m* cósmico
i polvere *f* interstellare
p poeira *f* cósmica
d kosmischer Staub *m*

654 cosmic radiation
f rayonnement *m* cosmique
e radiación *f* cósmica
i radiazione *f* cosmica
p radiação *f* cósmica
d kosmische Strahlung *f*

655 countercurrent
f contre-courant *m*
e contracorriente *f*
i controcorrente *f*
p contracorrente *f*
d Gegenstrom *m*

656 counterglow; antitwilight
f lueur *f* anticrépusculaire
e antihelio *m* crepuscular
i anticrepuscolo *m*
p zona *f* anticrepuscular
d Gegendämmerung *f*

657 covariation
f covariation *f*
e covariación *f*
i covariazione *f*
p co-variação *f*
d Kovariation *f*

658 cover
f couverture *f*
e cobertura *f*
i copertura *f*
p cobertura *f*
d Bedeckung *f*

659 crack
f fissure *f*
e grieta *f*
i fenditura *f*
p fenda *f*
d Bruch *m*

660 crater
f cratère *m*
e cráter *m*
i cratere *m*
p cratera *f*
d Krater *m*

661 creep; soil creep
f creep; reptation *f*
e deslizamiento *m*
i soliflussione *f*
p deslocamento *m*
d Kriechen *n*

662 crepuscular rays
f rayons *mpl* crépusculaires
e rayos *mpl* crepusculares
i raggi *mpl* crepuscolari
p raios *mpl* crepusculares
d Dämmerungstrahlen *mpl*

663 crescent moon
f lune *f* croissante
e luna *f* cresciente
i luna *f* crescente
p lua *f* crescente
d Mondsichel *f*

664 crevasse
f crevasse *f*
e grieta *f* del glaciar
i crevasse *f*
p criptoclase *f*
d Gletscherspalte *f*

665 critical frequency
f fréquence *f* critique
e frecuencia *f* crítica
i frequenza *f* critica
p frequência *f* crítica
d kritische Frequenz *f*; Grenzfrequenz *f*

666 critical humidity
f humidité *f* critique
e humedad *f* crítica
i umidità *f* critica
p umidade *f* crítica
d kritische Luftfeuchtigkeit *f*

667 critical temperature
f température *f* critique
e temperatura *f* crítica
i temperatura *f* critica
p temperatura *f* crítica
d kritische Temperatur *f*

668 critical velocity of wind
f vitesse *f* critique du vent
e velocidad *f* crítica del viento
i velocità *f* critica di vento
p velocidade *f* crítica do vento
d kritische Windgeschwindigkeit *f*

669 crop condition
f condition *f* de culture
e situación *f* de cultivo
i coltura *f* impiantata
p situação *f* da cultura
d Stand *m*; Kulturzustand *m*

670 crop data
f données *fpl* de production des végétaux
e datos *mpl* de producción de los vegetables
i dati *mpl* sulla produzione vegetale
p dados *mpl* sobre as culturas
d Anbaudaten *npl*

671 crop failure
f récolte *f* déficitaire
e pérdida *f* de cosecha
i cattivo raccolto *m*
p perda *f* da colheita
d Mißernte *f*

672 cropping advancement
f accélération *f* de récolte
e adelanto *m* de la cosecha
i anticipazione *f* della raccolta
p antecipação *f* da colheita
d Ernteverfrühung *f*

*** crop protection → 2074**

673 crop rotation
f rotation *f* des cultures
e rotación *f* de cultivos
i rotazione *f* colturale
p rotação *f* de culturas
d Fruchtwechsel *m*

674 cross currents
f courants *mpl* croisés
e corrientes *fpl* cruzadas
i correnti *fpl* incrociate
p correntes *fpl* cruzadas
d Kreuzströmungen *fpl*

675 cross section
f coupe *f* verticale
e corte *m* vertical
i sezione *f* trasversale
p corte *m* vertical
d Querschnitt *m*

676 cross wind; wind across; beam wind
f vent *m* de travers; vent *m* de côté
e viento *m* de través; viento *m* transversal
i vento *m* di fianco
p vento *m* de través
d Seitenwind *m*

677 crushed ice
f glace *f* concassée
e hielo *m* triturado
i ghiaccio *m* tritato
p gelo *m* triturado
d gemahlenes Eis *n*

678 crust
f croûte *f*
e crosta *f*
i crosta *f*
p crosta *f*
d Kruste *f*

679 cryoconite
f cryoconite *f*
e crioconita *f*
i crioconite *f*
p crioconita *f*
d Kryokonit *n*

680 cryohydrate
f cryohydrate *m*
e criohidrato *m*
i crioidrato *m*
p criohidrato *m*
d Kryohydrat *n*

681 cryology
(the scientific study of ice and snow)
f cryologie *f*
e criología *f*

i criologia *f*
p criologia *f*
d Kryologie *f*

682 cryopedology
f cryopédologie *f*
e criopedología *f*
i criopedologia *f*
p criopedologia *f*
d Kryopedologie *f*

683 cryopedometer
f cryopédomètre *m*
e criopedómetro *m*
i criopedometro *m*
p criopedômetro *m*
d Frostindikator *m*

684 cryosphere
f cryosphère *f*
e criosfera *f*
i criosfera *f*
p criosfera *f*
d Kryosphäre *f*

685 crystallization
f cristallisation *f*
e cristalización *f*
i cristallizzazione *f*
p cristalização *f*
d Kristallisation *f*

686 crystallization of water
f cristallisation *f* de l'eau
e cristalización *f* de la agua
i cristallizzazione *f* della acqua
p cristalização *f* da água
d Wasserkristallisation *f*

687 crystalloid
f cristalloïde *m*
e cristaloide *m*
i cristalloide *m*
p cristalóide *m*
d Kristalloid *n*

688 cube ice
f glace *f* en cubes
e hielo *m* en cubitos
i ghiaccio *m* in cubetti
p gelo *m* em cubos
d Würfeleis *n*

689 cultivated area; cultivated soil
f terre *f* cultivée
e tierra *f* cultivada; tierra *f* de cultivo
i arativo *m*; superficie *f* coltivata; terreno *m* coltivato
p terra *f* cultivada; solo *m* cultivado
d Kulturland *n*; Anbauland *n*; Kulturboden *m*

*** cultivated soil → 689**

690 cumuliform *adj*
(having the general shape of a cumulus cloud)
f en forme de cumulus
e cumuliforme
i cumuliforme
p em forma de cúmulo
d kumulusförmig

691 cumulonimbus
(heavy, dark cumulus with appearance of mountains or huge towers)
f cumulo-nimbus *m*
e cúmulonimbus *m*
i cumulonimbus *m*
p cúmulonimbus *m*
d Kumulonimbus *m*

692 cumulonimbus calvus
(highly-developed cirriform mass that produces frozen raindrops and flashes of lightning)
f cumulo-nimbus *m* calvus
e cúmulonimbus *m* calvus
i cumulonimbus *m* calvus
p cúmulonimbus *m* calvus
d Kumulonimbus *m* calvus

693 cumulonimbus incus; incus
(type of cloud characterized by an anvil-shaped top structure)
f cumulo-nimbus *m* incus; incus *m*
e cúmulonimbus *m* incus; incus *m*
i cumulonimbus *m* incus; incus *m*; incudine *f*
p cúmulonimbus *m* incus; frente *m* de trovoada
d Kumulonimbus *m* incus; amboßförmiger Kumulonimbus *m*

694 cumulus
(type of cloud characterized by elevation at 2000 feet with appearance of towers, mounds or domes)
f cumulus *m*
e cúmulo *m*
i cumulo *m*
p cúmulo *m*
d Kumulus *m*

695 cumulus congestus; congestus
(cumulus cloud characterized by a cauliflower appearance)
f cumulus *m* congestus; congestus *m*

e cúmulo *m* congestus; congestus *m*
i cumulo *m* congestus; congestus *m*
p cúmulo *m* congestus; congestus *m*
d Kumulus *m* congestus; Kongestus *m*

696 cumulus humilis
(cumulus cloud with a flattened appearance)
f cumulus *m* humilis
e cúmulo *m* humilis
i cumulo *m* humilis
p cúmulo *m* humilis
d Kumulus *m* humilis

697 cup anemometer
f anémomètre *m* à coupe
e anemómetro *m* de casquetes en cruz
i anemometro *m* a semisfere
p anemômetro *m* de copas cônicas
d Schalenkreuzwindmesser *m*

698 current; stream
f courant *m*
e corriente *f*
i corrente *f*
p corrente *f*
d Strom *m*

699 current chart
f carte *f* des courants
e carta *f* de las corrientes
i carta *f* delle correnti
p carta *f* das correntes
d Stromkarte *f*

700 current meter
f moulinet *m*
e molinete *m*
i mulinello *m*
p molinete *m*
d Wassermeßflügel *m*

701 current tables
f tables *mpl* des courants
e tablas *fpl* de las corrientes
i tavole *fpl* delle correnti
p tabelas *fpl* das correntes
d Strömungstafeln *fpl*

702 curvature of the earth
f courbure *f* de la terre
e curvatura *f* de la tierra
i curvatura *f* della terra
p curvatura *f* da terra
d Erdkrümmung *f*

703 cyanometer
(instrument used in meteorology to estimate the degrees of blueness of the sky)
f cyanomètre *m*
e cianómetro *m*
i cianometro *m*
p cianômetro *m*
d Zyanometer *n*

704 cyanosis
(discoloration of skin and mucous membranes due to deficient oxygenation in the blood)
f cyanose *f*
e cianosis *f*
i cianosi *f*
p cianose *f*
d Zyanose *f*

705 cycle
f cycle *m*
e ciclo *m*
i ciclo *m*
p ciclo *m*
d Zyklus *m*

706 cyclogenesis
(development of a cyclone)
f cyclogenèse *f*
e ciclogénesis *f*
i ciclogenesi *f*
p ciclogênese *f*
d Zyklogenesis *f*

707 cyclone
f cyclone *m*
e ciclón *m*
i ciclone *m*
p ciclone *m*
d Zyklon *m*

708 cyclone course; cyclone path
f route *f* de cyclone
e derrota *f* de ciclón
i rotta *f* di ciclone
p rota *f* de ciclone; curso *m* de ciclone
d Zyklonenweg *m*; Zyklonenbahn *f*

* **cyclone path → 708**

709 cyclone warning
f avertissement *m* de cyclone
e advertimiento *m* de ciclón
i avvertimento *m* di ciclone
p aviso *m* de ciclone
d Zyklonenwarnung *f*

710 cyclonic *adj*
f cyclonique
e ciclónico
i ciclonico
p ciclônico
d zyklonartig

* **cyclonic storm → 1401**

711 cyclonic system
f système *m* cyclonique
e sistema *m* ciclónico
i sistema *m* ciclonico
p sistema *m* ciclônico
d Zyklonwetterlage *f*

712 cyclostrophic force
f force *f* cyclostrophique
e fuerza *f* ciclostrófica
i forza *f* ciclostrofica
p força *f* ciclostrófica
d zyklostrophische Kraft *f*

713 cyclostrophic wind
f vent *m* cyclostrophique
e viento *m* ciclostrófico
i vento *m* ciclostrofico
p vento *m* ciclostrófico
d zyklostrophischer Wind *m*

D

714 dam
f barrage *m*
e represa *f*
i diga *f*
p represa *f*
d Damm *m*

* **dam** *v* → **1486**

715 damming of torrents
f correction *f* des torrents
e corrección *f* de torrentes
i regolazione *f* dei corsi d'acqua montani
p represamento *m* de correntes
d Wildbachverbauung *f*

* **damp** *adj* → **1851**

* **damp air** → **1852**

* **dampness** → **1396**

* **damp snow** → **2996**

716 damp weather
f temps *m* humide
e tiempo *m* húmedo
i tempo *m* umido
p tempo *m* úmido
d feuchtes Wetter *n*

717 dark *adj*
f sombre
e obscuro
i oscuro
p sombrio; escuro
d dunkel

718 dark hemisphere
f hémisphère *m* terrestre obscur
e hemisferio *m* terrestre obscuro
i emisfero *m* terrestre oscuro
p hemisfério *m* terrestre escuro
d unbeleuchtete Erdhälfte *f*

719 dark sky
f ciel *m* noir
e cielo *m* obscuro
i cielo *m* oscuro
p céu *m* escuro
d düsterer Himmel *m*

720 data
f données *fpl*
e datos *mpl*
i dati *mpl*
p dados *mpl*
d Daten *npl*

721 dating of glacier ice
f datation *f* de la glace
e datación *f* del hielo
i datazione *f* dei ghiaccio
p datação *f* do gelo
d Datierung *f* von Gletschereis

722 daughter product
f descendant *m* radioactif
e hijo *m* radiactivo
i figlio *m* radioattivo
p descendente *m* radioativo
d Tochterprodukt *n*

723 dawn; morning twilight
f aube *f*; crépuscule *m* du matin
e alba *f*; crepúsculo *m* matutino
i alba *f*; crepuscolo *m* mattutino
p aurora *f*; crepúsculo *m* matutino
d Morgendämmerung *f*

724 day
f jour *m*
e día *m*
i giorno *m*
p dia *m*
d Tag *m*

725 day airglow; day glow
f lueur *f* atmosphérique diurne
e resplandor *m* diurno de la atmósfera
i chiarore *m* diurno dell'atmosfera
p luz *f* celeste diurna
d Taghimmelsleuchten *n*

* **day glow** → **725**

726 daylength
f longueur *f* du jour
e longitud *f* del día
i durata *f* del giorno
p comprimento *m* do dia
d Tageslänge *f*

727 daylight
f lumière *f* diurne
e luz *f* diurna
i luce *f* diurna
p luz *f* do dia
d Tageslicht *n*

728 day of clear sky
f jour *m* de ciel clair
e día *m* despejado
i giorno *m* di cielo chiaro
p dia *m* de céu limpo
d heiterer Tag *m*

*** dead air region → 2245**

729 dead calm
f calme *m* plat
e calma *f* chicha
i calma *f* piatta
p calma *f* profunda
d Todesstille *m*

730 dead ice
f glace *f* morte
e glaciar *m* colgado
i ghiaccio *m* morto
p gelo *m* morto
d Toteis *n*

731 debacle
f débâcle *f*
e gran deshielo *m*
i disgelo *m*
p degelo *m*
d Eisbruch *m*

*** deblossoming → 745**

732 debris cone
f cône *m* couvert
e cono *m* de detrito
i cono *m* di detrito
p cone *m* de resíduos
d Sandkegel *m*

733 decade
f décennie *f*
e década *f*
i decade *f*
p década *f*
d Dekade *f*

734 December
f décembre *m*
e diciembre *m*
i dicembre *m*
p dezembro *m*
d Dezember *m*

735 decibar
(unit of pressure equivalent to 0.1 bar)
f décibar *m*
e decibar *m*
i decibar *m*
p decibar *m*
d Dezibar *n*

736 declination
f déclinaison *f*
e declinación *f*
i declinazione *f*
p declinação *f*
d Abweichung *f*; Deklination *f*

737 declination circle
f cercle *m* de déclinaison
e círculo *m* de declinación
i circolo *m* di declinazione
p círculo *m* de declinação
d Deklinationskreis *m*

738 decline; fall
f décline *f*
e decaimiento *m*
i decline *m*
p declínio *m*
d Fallen *n*

739 decrease in temperature; temperature drop; fall of temperature
f abaissement *m* de la température
e disminución *f* de la temperatura
i caduta *f* della temperatura
p queda *f* de temperatura
d Temperaturabnahme *f*

*** decrescent moon → 2923**

740 deepening of a depression
f creusement *m* d'une dépression
e intensificación *f* de una depresión
i approfondimento *m* di una depressione
p aprofundamento *m* de uma depressão
d Vertiefung *f* einer Zyklone

741 deepfreezing
f congélation *f* à basse température
e congelación *f* a baja temperatura
i congelazione *f* a bassa temperatura
p congelação *f* a baixa temperatura
d Tiefgefrieren *n*

742 deepfrozen fruit
f fruit *m* surgelé
e fruta *f* congelada
i frutta *f* surgelata
p fruta *f* congelada
d Tiefgefrierobst *n*

743 deepfrozen vegetables
f légumes *mpl* surgelés
e hortalizas *fpl* congeladas
i ortaggi *mpl* surgelati
p hortícolas *fpl* congeladas
d Tiefkühlgemüse *mpl*

744 deflation of snow
f déflation *f* de la neige
e deflación *f* de la nieve
i deflazione *f* della neve
p deflação *f* da neve
d Deflation *f* der Schneedecke

*** deflection of the plumb line → 787**

745 defloration; deblossoming
f défleuraison *f*
e desfloración *f*
i deflorazione *f*
p desfloração *f*
d Defloration *f*

746 deforestation
f déforestation *f*; déboisage *m*
e deforestación *f*
i deforestazione *f*
p desforestação *f*; desmatamento *m*
d Waldschlag *m*

747 deglaciation
f déglaciation *f*
e desglaciación *f*
i deglaciazione *f*
p deglaciação *f*
d Entgletscherung *f*

748 degradation of glaciers
f dégradation *f* des glaciers
e degradación *f* de glaciares
i degradazione *f* di ghiacciai
p degradação *f* de geleiras
d Gletscherrückzug *m*

749 degree
f degré *m*
e grado *m*
i grado *m*
p grau *m*
d Grad *m*

750 degree above zero
f degré *m* au-dessus de zéro
e grado *m* sobre cero
i grado *m* sopra zero
p grau *m* acima de zero
d Grad *m* über Null

751 degree below zero
f degré *m* au-dessous de zéro
e grado *m* bajo cero
i grado *m* sotto zero
p grau *m* abaixo de zero
d Grad *m* unter Null; Kältegrad *m*

752 degree Celsius
f degré *m* centigrade
e grado *m* centígrado
i grado *m* centigrado
p grau *m* centígrado
d Grad *m* Celsius; Zentigrad *m*

753 degree Fahrenheit
f degré *m* Fahrenheit
e grado *m* Fahrenheit
i grado *m* Fahrenheit
p grau *m* Fahrenheit
d Grad *m* Fahrenheit

754 degree of air saturation
f degré *m* de saturation en air
e grado *m* de saturación del aire
i grado *m* di saturazione dell'aria
p grau *m* de saturação do ar
d Luftsättigungsgrad *m*

755 degree of cloudiness
f degré *m* de nébulosité
e grado *m* de nebulosidad
i grado *m* di nebulosità
p grau *m* de nebulosidade
d Bewölkungsgrad *m*

756 degree of freedom
f degré *m* de liberté
e grado *m* de libertad
i grado *m* di libertà
p grau *m* de liberdade
d Freiheitsgrad *m*

757 degree of heat
f degré *m* de chaleur
e grado *m* de calor
i grado *m* di calore
p grau *m* de calor
d Hitzegrad *m*

758 degree of humidity; degree of wetness
f degré *m* d'humidité
e grado *m* de humedad
i grado *m* d'umidità
p grau *m* de umidade
d Feuchtigkeitsgrad *m*

759 degree of latitude
f degré *m* de latitude
e grado *m* de latitud
i grado *m* di latitudine
p grau *m* de latitude
d Breitengrad *m*

760 degree of saturation
f degré *m* de saturation
e grado *m* de saturación

i grado *m* di saturazione
p nível *m* de saturação
d Sättigungsgrad *m*

761 degree of temperature
f degré *m* de température
e grado *m* de temperatura
i grado *m* di temperatura
p grau *m* de temperatura
d Temperaturgrad *m*

*** degree of wetness → 758**

762 dehumidification
f déshumidification *f*
e deshumidificación *f*
i deumidificazione *f*
p desumidificação *f*
d Entfeuchtung *f*

763 dehydration
f déshydratation *f*
e deshidratación *f*
i desidratazione *f*
p desidratação *f*
d Dehydrierung *f*

764 de-icing air
f air *m* de dégivrage
e aire *m* caliente descongelador
i aria *f* calda antighiaccio
p ar *m* quente anti-gelo
d Warmluft *f*

765 delayed germination
f germination *f* retardée
e germinación *f* retardada
i germinazione *f* ritardata
p germinação *f* retardada
d Keimverzug *m*

766 delta
(alluvial deposit at the mouth of a river)
f delta *m*
e delta *m*
i delta *m*
p delta *m*
d Delta *n*

*** deluge → 2283**

767 dendrochronology
f dendrochronologie *f*
e dendrocronología *f*
i dendrocronologia *f*
p dendrocronologia *f*
d Dendrochronologie *f*

768 dendroclimatology
f dendroclimatologie *f*
e dendroclimatología *f*
i dendroclimatologia *f*
p dendroclimatologia *f*
d Dendroklimatologie *f*

769 dense smoke
f fumée *f* épaisse
e humareda *f*
i fumo *m* denso
p fumaça *f* densa
d Qualm *m*

770 density
f densité *f*
e densidad *f*
i densità *f*
p densidade *f*
d Dichte *f*

771 density of ice
f densité *f* de la glace
e densidad *f* del glaciar
i densità *f* di ghiaccio
p densidade *f* da geleira
d Eisdichte *f*

772 density of snow
f densité *f* de la neige
e densidad *f* de la nieve
i densità *f* della neve
p densidade *f* da neve
d Schneedichte *f*

773 density of the shadow
f densité *f* de l'ombre
e densidad *f* de la sombra
i densità *f* dell'ombra
p densidade *f* da sombra
d Schattendichte *f*

774 denudation
f dénudation *f*
e denudación *f*
i denudazione *f*
p desnudação *f*
d Denudation *f*

775 depletion
f tarissement *m*
e agotamiento *m*
i esaurimento *m*
p depleção *f*
d Entleerung *f*

*** depression → 839**

776 depression centre; eye of depression
f centre *m* de dépression
e centro *m* de depresión
i centro *m* di depressione
p centro *m* de uma depressão
d Depressionsmittelpunkt *m*

777 depression thunderstorm
f orage *m* de dépression
e tormenta *f* de depresión
i tempesta *f* di depressione
p temporal *m* de depressão
d Gewittersturm *m*

778 desalination
f dessalement *m*
e desalinización *f*
i dissalazione *f*
p dessalinização *f*
d Entsalzung *f*

*** descending current of air → 1021**

779 desert
f désert *m*
e desierto *m*
i deserto *m*
p deserto *m*
d Wüste *f*

780 desert dune
f dune *f* continentale
e duna *f* continental
i duna *f* continentale
p duna *f* continental
d Kontinentaldüne *f*

781 desertification
f désertification *f*
e desertización *f*
i desertificazione *f*
p desertificação *f*
d Wüstenbildung *f*

782 desert plateau
f plateau *m* désertique
e altiplano *m* desértico
i pianalto *m* desertico
p planalto *m* desértico
d Wüstentafel *f*

783 desert soil
f sol *m* désertique
e suelo *m* desértico
i terreno *m* desertico
p solo *m* desértico
d Wüstenboden *m*

784 deterioration of weather
f aggravation *f* du temps
e desmejoramiento *m* del tiempo
i peggioramento *m* del tempo
p empioramento *m* do tempo
d Wetterverschlechterung *f*

785 determination of gravity
f détermination *f* de l'intensité de la pesanteur
e determinación *f* de la gravedad
i determinazione *f* della gravità
p determinação *f* da gravidade
d Schweremessung *f*

786 development of water resources
f mise *f* en valeur des ressources en eau
e aprovechamiento *m* de los recursos hidráulicos
i valorizzazione *f* delle risorse idriche
p aproveitamento *m* dos recursos hídricos
d wasserwirtschaftliche Erschließung *f*

787 deviation of the vertical; deflection of the plumb line
f déviation *f* de la verticale
e desviación *f* vertical
i deviazione *f* della verticale
p desvio *m* da vertical
d Lotabweichung *f*

788 deviation of the wind
f déviation *f* du vent
e desviación *f* del viento
i deviazione *f* del vento
p desvio *m* dos ventos
d Abweichung *f* des Windes

789 dew
f rosée *f*
e rocío *m*
i rugiada *f*
p orvalho *m*
d Tau *m*

790 dew point; hoarfrost point; thawing point
f point *m* de rosée
e punto *m* del rocío; punto *m* de deshielo
i punto *m* di rugiada; punto *m* di disgelo
p ponto *m* de orvalho; ponto *m* de degelo
d Taupunkt *m*

791 dextrogyre *adj*
f dextrogyre
e dextrógiro
i destrogiro
p dextrógiro
d nach rechts ablenkend

792 diagnosis
f diagnostic *m*
e diagnosis *f*
i diagnosi *f*
p diagnose *f*
d Diagnose *f*

793 diamond dust
f poudrin *m* de glace
e polvillo *m* de hielo
i polvere *f* di diamante
p poeira *f* de diamante; prima *m* de gelo
d Diamantstaub *m*

794 diathermancy
f diathermansie *f*
e diatermancia *f*
i diatermano *m*
p diatermância *f*
d Diathermansie *f*

795 difference of temperature
f différence *f* de température
e diferencia *f* de temperatura
i differenza *f* di temperatura
p diferença *f* de temperatura
d Temperaturunterschied *m*

796 differential barometer
f baromètre *m* différentiel
e barómetro *m* diferencial
i barometro *m* differenziale
p barômetro *m* diferencial
d Differentialbarometer *n*

797 differential thermal analysis
f analyse *f* thermique différentielle
e análisis *f* térmica diferencial
i analisi *f* termo-differenziale
p análise *f* térmica diferencial
d Differentialthermoanalyse *f*

798 differentiation
f différentiation *f*
e diferenciación *f*
i differenziazione *f*
p diferenciação *f*
d Differenzierung *f*

799 diffraction
f diffraction *f*
e difracción *f*
i diffrazzione *f*
p difração *f*
d Diffraktion *f*

800 diffused lighting
f éclairage *m* diffus
e iluminación *f* indirecta
i luce *f* diffusa
p iluminação *f* difusa
d diffuse Beleuchtung *f*

801 diffuse luminous surface
f surface *f* lumineuse diffuse
e superficie *f* luminosa difusa
i velo *m* luminoso diffuso
p superfície *f* luminosa difusa
d diffus leuchtende Fläche *f*

802 diffusion
f diffusion *f*
e difusión *f*
i diffusione *f*
p difusão *f*
d Diffusion *f*

803 diffusive equilibrium
f équilibre *m* de diffusion
e equilibrio *m* de difusión
i equilibrio *m* di diffusione
p equilíbrio *m* de difusão
d Diffusionsgleichgewicht *n*

804 diluvial *adj*
f diluvial
e diluvial
i diluviale
p diluvial
d diluvial

805 diluvial soil
f sol *m* diluvial
e suelo *m* diluvial
i terreno *m* diluviale
p solo *m* diluvial
d Diluvialboden *m*

* **diluvium → 2283**

806 dip of horizon measurement error
f erreur *f* sur la mesure de la dépression de l'horizon
e error *m* de medida de la depresión del horizonte
i errore *m* di misura della depressione dell'orizzonte
p erro *m* de medida da depressão do horizonte
d Vermessungsfehler *m* der Kimmtiefe

807 dip of the horizon
f dépression *f* de l'horizon
e depresión *f* del horizonte
i depressione *f* dell'orizzonte
p depressão *f* do horizonte
d Kimmtiefe *f*

808 direction
f direction *f*
e dirección *f*
i direzione *f*
p direção *f*
d Richtung *f*

809 direct wave
f onde *f* directe
e onda *f* directa
i onda *f* diretta
p onda *f* direta
d direkte Welle *f*

810 discontinuity
f discontinuité *f*
e discontinuidad *f*
i discontinuità *f*
p descontinuidade *f*
d Diskontinuität *f*

811 dispersion
f dispersion *f*
e dispersión *f*
i dispersione *f*
p dispersão *f*
d Ausbreitung *f*

812 distance of the focus
f distance *f* du foyer
e distancia *f* del centro
i distanza *f* del centro
p distância *f* do centro
d Herdentfernung *f*

813 ditch
f fossé *m*
e foso *m*
i fosso *m*
p vala *f*
d Abzugsgraben *m*

814 diurnal *adj*
f diurne
e diurno
i diurno
p diurno
d täglich

815 diurnal changes
f changements *mpl* diurnes
e cambios *mpl* diurnos
i escursioni *fpl* diurne
p mudanças *fpl* diurnas
d Tagesänderungen *fpl*

816 diurnal inequality of tides
f inégalité *f* diurne des marées
e desigualdad *f* diurna de la mareas
i ineguaglianza *f* diurna della marea
p desigualdade *f* diurna das marés
d tägliche Ungleichheit *f* der Gezeiten

817 diurnal tide
f marée *f* diurne
e marea *f* diurnal
i marea *f* diurna
p maré *f* diurna
d tägliche Gezeiten *fpl*

818 diurnal variation
f variation *f* diurne
e variación *f* diurna
i variazione *f* diurna
p variação *f* diurna
d Tagesschwankung *f*

819 diurnal wave
f onde *f* diurne
e onda *f* diurna
i onda *f* diurna
p onda *f* diurna
d tägliche Welle *f*

820 divergence
f divergence *f*
e divergencia *f*
i divergenza *f*
p divergência *f*
d Divergenz *f*

821 dog days
f canicule *f*
e canícula *f*
i canicola *f*
p canícula *f*
d Hundstage *mpl*

*** doldrums → 1520**

822 dormant volcano
f volcan *m* inactif
e volcán *m* inactivo
i vulcano *m* inattivo
p vulcão *m* inativo
d untätiger Vulkan *m*

823 dosimeter
f dosimètre *m*
e dosímetro *m*
i dosimetro *m*
p dosímetro *m*
d Dosimeter *n*

824 double cropping
f culture *f* double
e cultivo *m* hurtado
i coltura *f* furtiva

p cultivo *m* de colheita dupla
d Zweifachnützung *f*

825 down current
f vent *m* descendant
e viento *m* descendente
i vento *m* discendente
p vento *m* descendente
d Abwind *m*

826 downstream *adj*
f en aval
e río abajo
i a valle
p vazante
d stromabwärts

827 down wind; head wind
f vent *m* longitudinal; vent *m* de bout; vent *m* d'arrière
e viento *m* contrario
i vento *m* frontale; vento *m* en poppa
p vento *m* contrário
d Längswind *m*

828 drainage
f drainage *m*
e drenaje *f*
i drenaggio *m*
p drenagem *f*
d Entwässerung *f*

829 drainage area; drainage basin; catchment area
f bassin *m* hydrographique
e cuenca *f* hidrográfica
i bacino *m* di drenaggio
p bacia *f* hidrográfica
d Wassereinzugsgebiet *n*

* **drainage basin → 829**

830 drainage of glaciers
f drainage *m* des glaciers
e drenaje *f* de los glaciers
i drenaggio *m* di ghiacciai
p drenagem *f* das geleiras
d Entwässerung *f* der Gletscher

831 drift
f dérive *f*
e deriva *f*
i deriva *f*
p deriva *f*
d Trift *f*

832 drift current
f courant *m* causé par le vent
e corriente *f* causada por el viento
i corrente *f* causata dal vento
p corrente *f* de deriva
d Triftströmung *f*

833 drift ice chart
f carte *f* de glaces en dérive
e carta *f* de carámbano a la deriva
i carta *f* di ghiacci alla deriva
p carta *f* de gelo à deriva
d Treibeiskarte *f*

834 drifting snow
f chasse-neige *f* basse
e borrasca *f* de nieve
i neve *f* polvere
p neve *f* levantada pelo vento
d Staubschnee *m*

* **drift of continents → 627**

835 drinking water
f eau *f* potable
e agua *f* potable
i acqua *f* potabile
p água *f* potável
d Trinkwasser *n*

836 drinking water supply
f approvisionnement *m* en eau potable
e abastecimento *m* de agua potable
i approvvigionamento *m* di acqua potabile
p abastecimento *m* de água potável
d Trinkwasserversorgung *f*

837 driven to leeward
f déventé
e sotaventado
i sottoventato
p sotaventeado
d bekalmt

838 drizzling rain
f pluie *f* ruisselante
e llovizna *f*
i puiggerella *f*
p névoa *f* úmida
d Sprühregen *m*

839 drop in pressure; depression
f dépression *f*
e depresión *f*
i depressione *f*
p depressão *f*
d Tiefdruck *m*

* **droplet → 1659**

840 drosometer
f drosomètre *m*
e drosómetro *m*
i drosometro *m*
p drosômetro *m*
d Taumesser *m*

*** drought → 855**

841 dry *adj*
f sec
e seco
i secco
p seco
d trocken

842 dry adiabatic
f adiabatique *m* sec
e adiabata *f* seca
i adiabata *f* secca
p adiabático *m* seco
d trockene Adiabate *f*

*** dry-adiabatic lapse rate → 35**

843 dry barometer
f baromètre *m* sec
e barómetro *m* seco
i barometro *m* secco
p barômetro *m* seco
d Trockenbarometer *n*

844 dry bulb thermometer
f thermomètre *m* à boule sèche
e termómetro *m* de bola seca
i termometro *m* a bulbo secco
p termômetro *m* de bulbo seco
d Trockenthermometer *n*

845 dry cold air
f air *m* sec et froid
e aire *m* seco y frío
i aria *f* secca e fredda
p ar *m* seco e frio
d trockene kalte Luft *f*

846 dry farming; dryland farming
f culture *f* en sols arides
e cultivo *m* de secano
i aridocoltura *f*
p cultura *f* em solos áridos
d Trockenfeldbau *m*

847 dry fog
f brouillard *m* sec
e niebla *f* seca
i nebbia *f* secca
p névoa *f* seca
d trockener Nebel *m*

848 dry heat
f chaleur *f* sèche
e calor *m* seco
i calore *m* secco
p calor *m* seco
d Trockenwärme *f*

849 dry ice
f glace *f* sèche
e hielo *m* seco
i ghiaccio *m* secco
p gelo *m* seco
d Trockeneis *n*

850 drying floor
f plancher *m* de séchage
e secadero *m*
i suolo *m* al essiccazione
p chão *m* de secagem
d Trocknungsraum *m*

*** dryland farming → 846**

851 dry matter
f matière *f* sèche
e materia *f* seca
i materia *f* secca
p matéria *f* seca
d Trockensubstanz *f*

*** dryness → 855**

*** dry season → 855**

852 dry snow
f neige *f* sèche
e nieve *f* seca
i neve *f* secca
p neve *f* seca
d trockener Schnee *m*

853 dry snow avalanche
f avalanche *f* de neige sèche
e avalancha *f* de nieve seca
i valanga *f* di neve secca
p avalancha *f* de neve seca
d Trockenschneelawine *f*

854 dry snow line
f ligne *f* de neiges sèches
e línea *f* de nieves secas
i linea *f* di nevi secca
p linha *f* de neves secas
d Trockenschneegrenze *f*

855 dry spell; dry season; low water; drought; aridity; dryness
f période *f* sèche; sécheresse *f*; étiage *m*
e período *m* seco; sequía *f*; aridez *f*; seca *f*

i periodo *m* di siccità; stagione *f* secca; siccità *f*
p período *m* de seca; estiagem *f*; seca *f*
d Trockenperiode *f*; Dürre *f*; Trockenheit *f*

856 dry warm air
f air *m* sec et chaud
e aire *m* seco y caliente
i aria *f* secca e calda
p ar *m* seco e quente
d trockene warme Luft *f*

857 dry weather
f temps *m* sec
e tiempo *m* seco
i tempo *m* secco
p tempo *m* seco
d trockenes Wetter *n*

858 dune
(mound or ridge of sand formed by the wind)
f dune *f*
e duna *f*
i duna *f*
p duna *f*
d Düne *f*

859 duration of leaf wetness
f durée *f* d'humectation
e duración *f* de humedad foliar
i tempo *m* di bagnatura fogliare
p duração *f* da umidade foliar
d Benetzungsdauer *f*

860 duration of sunshine
f durée *f* de l'insolation
e horas *fpl* de insolación
i durata *f* dello splendore del sole
p duração *f* do brilho do sol
d Sonnenscheindauer *f*

* **dusk → 2843**

861 dust
f poussière *f*
e polvo *m*
i pulviscolo *m*
p poeira *f*
d Staub *m*

862 dust avalanche
f avalanche *f* sèche
e alud *m* seco
i valanga *f* di cenere
p avalancha *f* de cinzas
d vulkanische Staublawine *f*

863 dust cloud
f nuage *m* de poussière
e nube *f* de polvo
i nube *f* di polvere
p nuvem *f* de pó
d Staubwolke *f*

864 dust counter
f compteur *m* de poussière; pulvimètre *m*
e pulvioscopio *m*
i contatore *m* dei granelli di polvere
p polvoscópio *m*
d Staubzähler *m*

* **dust devil → 2301**

* **duster → 866**

865 dust laden air; air permeated with dust
f air *m* poussiéreux
e aire *m* polvoriento
i aria *f* polverosa
p ar *m* carregado de poeira; ar *m* poeirento
d Staubluft *f*

866 dust storm; duster
f tempête *f* de poussière
e tempestad *f* de polvo
i tempesta *f* di polvere
p tempestade *f* de poeira
d Staubsturm *m*

867 dynamic cooling
f refroidissement *m* dynamique
e enfriamiento *m* dinámico
i raffreddamento *m* dinamico
p resfriamento *m* dinâmico
d dynamische Abkühlung *f*

868 dynamic glaciology; mechanics of ice
f glaciologie *f* dynamique; mécanique *f* de la glace
e glaciología *f* dinámica
i glaciologia *f* dinamica; meccanica *f* del ghiaccio
p glaciologia *f* dinâmica
d dynamische Glaziologie *f*; Eismechanik *f*

869 dynamic meteorology
f météorologie *f* dynamique
e meteorología *f* dinámica
i meteorologia *f* dinamica
p meteorologia *f* dinâmica
d dynamische Meteorologie *f*

870 dynamic pressure
f pression *f* dynamique
e presión *f* dinámica
i pressione *f* dinamica
p pressão *f* dinâmica
d dynamischer Druck *m*

871 dynamic stability
f stabilité *f* dynamique
e estabilidad *f* dinámica
i stabilità *f* dinamica
p estabilidade *f* dinâmica
d dynamische Stabilität *f*

872 dynamometamorphism of ice
f dynamométamorphisme *m* de la glace
e metamorfismo *m* dinámico del hielo
i metamorfismo *m* dinamico del ghiaccio
p metamorfismo *m* dinâmico do gelo
d dynamische Metamorphose *f* von Eis

E

873 early crop
f récolte *f* précoce
e cosecha *f* temprana; cosecha *f* anticipada
i raccolta *f* precoce
p colheita *f* prematura
d Früherernte *f*

874 early frost
f gelée *f* précoce
e helada *f* precoz
i gelata *f* precoce
p geada *f* precoce
d Frühfrost *m*

*** earth → 2437**

875 earth current
f courant *m* tellurique
e corriente *f* telúrica
i corrente *f* terrestre
p corrente *f* telúrica
d Erdstrom *m*

876 earth ellipsoid
f ellipsoïde *m* terrestre
e elipsoide *m* terrestre
i ellissoide *m* terrestre
p elipsóide *m* terrestre
d Erdellipsoid *n*

*** earthflow → 1613**

877 earthquake; earth tremor; seism
f tremblement *m* de terre; séisme *m*
e temblor *m* de tierra; sismo *m*
i terremoto *m*; sismo *m*
p terremoto *m*; sismo *m*
d Erdbeben *n*

878 earthquake focus
f foyer *m* du tremblement de terre
e foco *m* del terremoto
i focus *m* dei terremoti
p foco *m* do terremoto
d Herd *m* des Erdbebens

*** earth's atmosphere → 232**

879 earth's charge
f charge *f* terrestre
e carga *f* terrestre
i carica *f* terrestre
p carga *f* terrestre
d Erdladung *f*

880 earth's core
f noyau *m* de la terre
e núcleo *m* central de la tierra
i nucleo *m* terrestre
p núcleo *m* central da terra
d Erdkern *m*

881 earth's crust
f croûte *f* terrestre
e crosta *f* terrestre
i crosta *f* terrestre
p crosta *f* terrestre
d Erdkruste *f*

882 earth's field
f champ *m* terrestre
e campo *m* terrestre
i campo *m* terrestre
p campo *m* terrestre
d Erdfeld *n*

883 earth's interior
f intérieur *m* de la terre
e interior *m* de la tierra
i interno *m* della terra
p interior *m* da terra
d Erdinneres *n*

884 earth's magnetic field
f champ *m* magnétique terrestre
e campo *m* magnético terrestre
i campo *m* magnetico terrestre
p campo *m* magnético terrestre
d erdmagnetisches Feld *n*

885 earth's magnetic pole; magnetic pole of the earth
f pôle *m* magnétique terrestre
e polo *m* magnético terrestre
i polo *m* magnetico terrestre
p pólo *m* magnético terrestre
d magnetischer Erdpol *m*

886 earth's magnetism; terrestrial magnetism
f magnétisme *m* terrestre
e magnetismo *m* terrestre
i magnetismo *m* terrestre
p magnetismo *m* terrestre
d Erdmagnetismus *m*

887 earth's mantle
f manteau *m* terrestre
e manto *m* de la tierra
i manto *m* terrestre
p manto *m* da terra
d Mantel *m*; Erdmantel *m*

*** earth's oblateness → 1953**

888 earth spheroid
f sphéroïde *m* terrestre
e esferoide *m* terrestre
i sferoide *m* terrestre
p esferóide *m* terrestre
d Erdsphäroid *n*

889 earth's shadow
f ombre *f* de la terre
e sombra *f* de la tierra
i ombra *f* della terra
p sombra *f* da terra
d Erdschatten *m*

890 earth's surface
f surface *f* de la terre
e superficie *f* de la tierra
i superficie *f* della terra
p superfície *f* da terra
d Erdoberfläche *f*

* **earth thermometer → 1205**

* **earth tremor → 877**

891 earth water
f eau *f* tellurique
e agua *f* telúrica
i acqua *f* tellurica
p água *f* telúrica
d tellurisches Wasser *n*

892 east
f est *m*
e este *m*
i est *m*
p leste *m*
d Osten *m*

893 East-Australian Current
f Courant *m* australien oriental
e Corriente *f* australiana oriental
i Corrente *f* australiana orientale
p Corrente *f* australiana oriental
d Ostaustralstrom *m*

894 east coast fever; Rhodesian fever
f fièvre *f* de la côte australe
e fiebre *f* de la costa oriental de Africa
i febbre *f* della costa dell'Africa orientale
p febre *f* da costa oriental da Africa
d ostafrikanisches Küstenfieber *n*; rhodesisches Fieber *n*

895 Eastern Mediterranean
f Méditerrannée *f* orientale
e Mediterráneo *m* oriental
i Mediterraneo *m* orientale
p Mediterrâneo *m* oriental
d östliches Mittelmeer *n*

896 eastern wind
f vent *m* d'est
e viento *m* del este
i vento *m* d'este
p vento *m* leste
d Ostwind *m*

897 East-European time
f heure *f* de l'Europe orientale
e hora *f* de la Europa oriental
i ora *f* dell'Europa orientale
p hora *f* da Europa oriental
d osteuropäische Zeit *f*

898 ebb; ebb tide; falling tide
f marée *f* descendante; jusant *m*; reflux *m*
e reflujo *m*; marea *f* vaciante; marea *f* descendente
i marea *f* decrescente; riflusso *m*
p maré *f* vazante; ressaca *f*; vazante *f*
d Ebbe *f*

* **ebb tide → 898**

899 ebb tide and flood tide
f flux *m* et reflux *m*
e flujo *m* y reflujo *m*
i flusso *m* e riflusso *m*
p fluxo *m* e refluxo *m*
d Ebbe *f* und Flut *f*

* **ebullition → 406**

900 echo
f écho *m*
e eco *m*
i eco *m*
p eco *m*
d Echo *n*

901 eclipse
f éclipse *f*
e eclipse *m*
i eclisse *f*
p eclipse *m*
d Finsternis *f*

902 ecliptic *adj*
f écliptique
e eclíptico
i eclittico
p eclíptico
d ekliptisch

903 ecobiotic adaptation
f adaptation *f* écobiotique
e adaptación *f* ecobiótica
i adattamento *m* ecobiotico
p adaptação *f* ecobiótica
d ökobiotische Anpassung *f*

904 ecological *adj*
f écologique
e ecológico
i ecologico
p ecológico
d ökologisch

905 ecological engineering
f génie *m* écologique
e ingeniería *f* ecológica
i ingegneria *f* ecologica
p engenharia *f* ecológica
d Natur- und Landschaftgestaltung *f*

906 ecological niche
f niche *f* écologique
e nicho *m* ecológico
i nicchia *f* ecologica
p nicho *m* ecológico
d ökologische Nische *f*

907 ecology
f écologie *f*
e ecología *f*
i ecologia *f*
p ecologia *f*
d Ökologie *f*

908 economic injury level
f seuil *m* de nocivité économique
e nivel *m* económico de daño
i limite *m* economico di aggressione
p nível *m* econômico de dano
d wirtschaftliche Schadensschwelle *f*

909 ecosphere
f écosphère *f*
e ecosfera *f*
i ecosfera *f*
p ecosfera *f*
d Ökosphäre *f*

910 ecosystem
f écosystème *m*
e ecosistema *m*
i ecosistema *m*
p ecossistema *m*
d Ökosystem *n*

*** eddying current → 912**

*** eddying wind → 1587**

*** eddy motion → 2920**

911 eddy velocity
f vitesse *f* turbulente
e velocidad *f* turbulenta
i velocità *f* turbolenta
p velocidade *f* turbilhonar
d mittlere Geschwindigkeitsschwankung *f*

912 eddy water; eddying current; whirlpool; vortex
f tourbillon *m*
e remolino *m*; vórtice *m*
i gorgo *m*; vortice *m*
p redemoinho *m*; voragem *f*
d Strudel *m*; Stromwirbel *m*

913 edged gust
f rafale *m* aigu
e ráfaga *f* aguda
i raffica *f* acuta
p ráfaga *f* aguda
d scharfer Windstoß *m*

914 effective temperature
f température *f* effective
e temperatura *f* efectiva
i temperatura *f* effettiva
p temperatura *f* efetiva
d Strahlungstemperatur *f*

915 efflorescence
f effloraison *f*; efflorescence *f*
e eflorescencia *f*
i efflorescenza *f*
p eflorescência *f*
d Effloreszenz *f*

916 effusion
f effusion *f*
e efusión *f*
i effusione *f*
p efusão *f*
d Effusion *f*

917 eighth
f octave *f*
e octavo *m*
i ottavo *m*
p oitavo *m*
d Achtel *n*

918 elasticity
f élasticité *f*
e elasticidad *f*
i elasticità *f*
p elasticidade *f*
d Elastizität *f*

919 electric anemograph
f anémographe *m* électrique
e anemógrafo *m* eléctrico
i anemografo *m* elettrico
p anemógrafo *m* elétrico
d elektrischer Anemograph *m*

920 electric conductivity
f conductibilité *f* électrique
e conductibilidad *f* eléctrica
i conducibilità *f* elettrica
p condutibilidade *f* elétrica
d elektrische Leitfähigkeit *f*

921 electric storm
f tempête *f* électrique
e tempestad *f* eléctrica
i temporale *m* elettrico
p tempestade *f* elétrica
d elektrischer Sturm *m*

922 electric thermometer
f thermomètre *m* électrique
e termómetro *m* eléctrico
i termometro *m* elettrico
p termômetro *m* elétrico
d Elektrothermometer *n*

923 electron density
f densité *f* électronique
e densidad *f* electrónica
i densità *f* elettronica
p densidade *f* eletrônica
d Elektronendichte *f*

924 elevation
f élévation *f*
e elevación *f*
i elevazione *f*
p elevação *f*
d Meereshöhe *f*

925 El Niño
(warm ocean current setting along the coast of Ecuador and Peru at Christmas time)
f El Niño *m*
e El Niño *m*
i El Niño *m*
p El Niño *m*
d El Niño *m*

* **eluviation → 1630**

926 embata
(southwest wind on the Canary Islands)
f embata *m*
e embata *m*
i embata *m*
p embata *m*
d Embata *m*

* **emission of lava → 1626**

927 emissivity
f émissivité *f*
e emisividad *f*
i emissività *f*
p emissividade *f*
d Emissionskraft *f*

928 endemic grippe
f grippe *f* endémique
e gripe *f* endémica
i grippe *f* endemica
p gripe *f* endêmica
d endemische Grippe *f*

929 endogenous *adj*
f endogène
e endógeno
i endogeno
p endógeno
d endogen

930 engineering glaciology
f génie *m* glaciologique
e ingeniería *f* glaciológica
i ingegneria *f* glaciologica
p engenharia *f* glaciológica
d Ingenieurwesenglaziologie *f*

931 engineering ice
f glace *f* technique
e hielo *m* técnico
i ghiaccio *m* tecnico
p gelo *m* técnico
d technisches Eis *n*

932 enthalpy
f enthalpie *f*
e entalpia *f*
i entalpia *f*
p entalpia *f*
d Enthalpie *f*

933 entrainment
f entraînement *m*
e arrastre *m*
i intrusione *f*
p arrasto *m* convectivo
d Mitreißen *n*

934 entropy
f entropie *f*
e entropía *f*
i entropia *f*
p entropia *f*
d Entropie *f*

935 environment
f environnement
e medio *m* ambiente
i mezzo *m* ambiente
p meio *m* ambiente
d Umwelt *f*

936 environmental conditions
f conditions *fpl* du milieu
e condiciones *fpl* ambientales
i condizioni *fpl* ambientali
p condições *fpl* ambientais
d Umweltbedingungen *fpl*

* **environmental conservation → 945**

937 environmental factor
f facteur *m* du milieu
e factor *m* ambiental
i fattore *m* ambientale
p fator *m* do meio ambiente
d Umweltfaktor *m*

938 environmental geology
f géologie *f* de l'environnement
e geología *f* del medio ambiente
i geologia *f* ambientale
p geologia *f* ambiental
d Umweltgeologie *f*

939 environmental impact assessment
f étude *f* d'impact d'environnement
e estudio *m* de impacto ambiental
i determinazione *f* di impatto ambientale
p estudo *m* do impacto ambiental
d Umweltstudie *f*

940 environmentalist
f spécialiste *m* de l'environnement
e especialista *m* en medio ambiente
i ambientalista *m*
p ambientalista *m*
d Umweltspezialist *m*

941 environmentally compatible
f compatible avec l'environnement
e compatible con el medio ambiente
i compatibile sotto il profilo ambientale
p compatível com o meio ambiente
d umweltverträglich

942 environmental medicine
f médecine *f* des milieux
e medicina *f* ambiental
i medicina *f* dell'ambiente
p medicina *f* ambiental
d Milieumedizin *f*

943 environmental plan
f plan *m* d'aménagement rural
e plan *m* paisajístico
i pianificazione *f* ambientale
p plano *m* ambiental
d Plan *m* für Landschaftsschutz und Entwicklung

944 environmental pollution
f contamination *f* ambientale
e contaminación *f* ambiental
i inquinamento *m* dell'ambiente
p poluição *f* do meio ambiente
d Umweltverseuchung *f*

945 environmental protection; environmental conservation
f protection *f* de l'environnement
e protección *f* del medio ambiente
i protezione *f* dell'ambiente
p proteção *f* do meio ambiente
d Umweltschutz *m*

946 environmental theory
f théorie *f* des milieux
e teoría *f* del medio ambiente
i teoria *f* dell'ambiente
p teoria *f* do meio ambiente
d Milieutheorie *f*

947 eolian *adj*; **aeolian** *adj*
f éolien
e eólico
i eolico
p eólico
d äolisch; Wind...

948 eolian feature
f morphologie *f* éolienne
e morfología *f* eólica
i morfologia *f* eolica
p morfologia *f* eólica
d äolische Morphologie *f*

* **eolian sediment → 3031**

949 eolian vibration
f vibration éolienne
e vibración *f* eólica
i vibrazione *f* eolica
p vibração *f* eólica
d winderregte Schwingung *f*

950 epicentre
f épicentre *m*
e epicentro *m*
i epicentro *m*
p epicentro *m*
d Epizentrum *n*

951 epidemic grippe
f grippe *f* epidémique
e gripe *f* epidémica
i grippe *f* epidemica
p gripe *f* epidêmica
d epidemische *f* Grippe

952 equation of state
f équation *f* d'état
e ecuación *f* de estado
i equazione *f* di stato
p equação *f* de estado
d Zustandsgleichung *f*

953 equation of time
f équation *f* du temps
e ecuación *f* de tiempo
i equazione *f* del tempo
p equação *f* do tempo
d Zeitgleichung *f*

954 equations of motion
f équations *fpl* du mouvement
e ecuaciones *fpl* de movimiento
i equazioni *fpl* di moto
p equações *fpl* de movimento
d Bewegungsgleichungen *fpl*

955 equator
f équateur *m*
e ecuador *m*
i equatore *m*
p equador *m*
d Äquator *m*

956 equatorial air
f air *m* équatorial
e aire *m* ecuatorial
i aria *f* equatoriale
p ar *m* equatorial
d Äquatorialluft *f*

957 equatorial climate
f climat *m* équatorial
e clima *m* ecuatorial
i clima *m* equatoriale
p clima *m* equatorial
d äquatorialisches Klima *n*

958 equatorial coordinates
f coordonnées *fpl* équatoriales
e coordenadas *fpl* ecuatoriales
i coordinate *fpl* equatoriali
p coordenadas *fpl* equatoriais
d Äquatorialkoordinaten *fpl*

959 equatorial countercurrent
f contre-courant *m* équatorial
e contracorriente *f* ecuatorial
i controcorrente *f* equatoriale
p contracorrente *f* equatorial
d Äquatorialgegenströmung *f*

960 equatorial current
f courant *m* équatorial
e corriente *f* ecuatorial
i corrente *f* equatoriale
p corrente *f* equatorial
d Äquatorialstrom *m*

961 equatorial parallax on the horizon
f parallaxe *f* horizontale équatoriale
e paralaje *f* horizontal ecuatorial
i parallasse *f* orizzontale equatoriale
p paralaxe *f* horizontal equatorial
d äquatoriale Horizontalparallaxe *f*

962 equatorial plane
f plan *m* de l'équateur terrestre
e plano *m* del ecuador terrestre
i piano *m* dell'equatore terrestre
p plano *m* do equador terrestre
d Ebene *f* des Erdäquators

963 equatorial tide
f marée *f* équatoriale
e marea *f* ecuatorial
i marea *f* equatoriale
p maré *f* equatorial
d Äquatorialgezeiten *fpl*

964 equatorial trough
f région *f* des calmes équatoriaux
e zona *f* de calmas ecuatoriales
i saccatura *f* equatoriale
p zona *f* de calmas equatoriais
d äquatoriale Kalmengürtel *m*

965 equatorial zone
f zone *f* équatoriale
e zona *f* ecuatorial
i ambiente *m* equatoriale
p zona *f* equatorial
d Äquatorialzone *f*

966 equinoctial point
f point *m* équinoxial
e punto *m* equinoccial
i punto *m* equinoziale
p ponto *m* equinocial
d Äquinoktialpunkt *m*

967 equinoctial spring tide
f marée *f* d'équinoxe de printemps
e marea *f* equinoccial de primavera
i marea *f* equinoziale di primavera
p maré *f* equinocial de primavera
d Springtide *f* zur Zeit der Tagundnachtgleiche

968 equinoctial storm
f coup *m* de vent de l'équinoxe
e temporal *m* del equinoccio
i tempeste *f* equinoziale
p tempestade *f* do equinócio
d Äquinoktialstürm *m*

969 equinoctial tide; spring tide
f marée *f* d'équinoxe; marée *f* de vive-eau
e marea *f* equinoccial; marea *f* viva
i marea *f* equinoziale
p maré *f* equinocial; maré *f* de águas vivas
d Äquinoktialgezeiten *fpl*; Springtide *f*

970 equinox
(either of the two times each year when the sun crosses the equator)
f équinoxe *m*
e equinoccio *m*
i equinozio *m*
p equinócio *m*
d Äquinoktium *n*

971 erg
(unit of energy or work in the centimeter-gram-second system)
f erg *m*
e ergio *m*
i erg *m*
p erg *m*
d Erg *n*

972 eroded soil
f sol *m* érodé
e suelo *m* erosionado
i terreno *m* eroso
p solo *m* erosivo
d erodierter Boden *m*

973 erosion; soil loosening
f érosion *f*; ameublissement *m* du sol
e erosión *f*; mullimiento *m* del suelo
i erosione *f*; dissodamento *m* del terreno
p erosão *f*; enfraquecimento *m* do solo
d Erosion *f*; Bodenabtrag *f*; Lockerung *f* des Bodens

974 error
f erreur *f*
e error *m*
i errore *m*
p erro *m*
d Fehler *m*

* **error in height → 116**

975 error in latitude
f erreur *f* en latitude
e error *m* de latitud
i errore *m* in latitudine
p erro *m* de latitude
d Breitenfehler *m*

976 error in longitude
f erreur *f* en longitude
e error *m* de longitud
i errore *m* di longitudine
p erro *m* de longitude
d Längenfehler *m*

977 ertor
(effective temperature of ozone layer around the earth)
f ertor *m*
e ertor *m*
i ertor *m*
p ertor *m*
d Ertor *m*

978 eruption
f éruption *f*
e erupción *f*
i eruzione *f*
p erupção *f*
d Ausbruch *m*

979 eruption of Krakatoa
f éruption *f* du Krakatoa
e explosión *f* del Krakatoa
i eruzione *f* del Krakatoa
p explosão *f* do Krakatoa
d Krakatauausbruch *m*

980 escape of gases
f libération *f* des gaz
e escape *m* de gases
i fuga *f* dei gas
p escape *m* de gases
d Entweichen *n* von Gasen

981 estivation
f estivation *f*
e veraneo *m*
i estivazione *f*
p estivação *f*
d Ästivation *f*

982 estuarine facies
f faciès *m* estuarin
e faz *f* lagunar
i faccia *f* lagunare
p fácies *f* lagunar
d lagunäres Fazies *n*

983 estuary
f estuaire *m*
e estuario *m*
i estuario *m*
p estuário *m*
d Ästuarium *n*

984 Etesian winds
f vents *mpl* étésiens
e vientos *mpl* etesios
i venti *mpl* estivi
p etésios *mpl*
d Etesienwinde *mpl*

985 evaporate *v*
f évaporer
e evaporar
i evaporare
p evaporar
d verdunsten

986 evaporation
f évaporation *f*
e evaporación *f*
i evaporazione *f*
p evaporação *f*
d Verdunstung *f*

987 evaporative ice
f givre *m* d'évaporation
e hielo *m* de evaporación
i ghiaccio *m* di evaporazione
p gelo *m* de evaporação
d Verdunstungseis *n*

988 evaporimeter; atmidometer
f évaporimètre *m*; atmidomètre *m*
e evaporímetro *m*; atmidómetro *m*
i evaporimetro *m*; atmidometro *m*
p evaporímetro *m*; atmidômetro *m*
d Evaporimeter *n*

989 evapotranspiration
f évapotranspiration *f*
e evapotranspiración *f*
i evapotraspirazione *f*
p evapotranspiração *f*
d Evapotranspiration *f*

990 evening primrose
f onagre *f*
e enotera *f*
i oenothera *f*
p primavera *f* da tarde
d Nachtkerze *f*

991 examination of the soil
f sondage *m* du sol
e sondeo *m* del terreno
i sondaggio *m* del terreno
p sondagem *f* do terreno
d Bodenuntersuchung *f*

992 excessive heat
f chaleur *f* excessive
e calor *m* excesivo
i caldo *m* eccessivo
p calor *m* excessivo
d übermäßige Hitze *f*

993 excess of heat
f excès *m* de chaleur
e exceso *m* de calor
i eccesso *m* di calore
p excesso *m* de calor
d Wärmeüberschuß *m*

994 excitation mechanism
f mécanisme *m* d'excitation
e mecanismo *m* de excitación
i meccanismo *m* di eccitazione
p mecanismo *m* de excitação
d Anregungsprozeß *m*

995 exhaust trail
f traînée *f* d'échappement
e estela *f* de escape
i scia *f* di scarico
p trilha *f* de escapamento
d Kondensationsspur *f*

996 exogenous *adj*
f exogène
e exógeno
i esogeno
p exógeno
d exogen

997 exosphere; spray region; fringe region
f exosphère *f*; passage *m* vers l'espace interstellaire
e exosfera *f*; margen *f* de la zona
i esosfera *f*; atmosfera *f* esterna
p exosfera *f*; região *f* de atomização
d Exosphäre *f*; äußere Atmosphäre *f*

998 expansion
f détente *f*
e dilatación *f*
i espansione *f*
p expansão *f*
d Ausdehnung *f*

999 explosion cloud
f fumée *f* d'éclatement
e nube *f* de explosión
i fumo *m* dello scopio
p nuvem *f* de explosão
d Sprengwolke *f*

1000 explosion wave
f onde *f* d'explosion
e onda *f* de explosión
i onda *f* d'esplosione
p onda *f* de explosão
d Explosionswelle *f*

1001 exposure
f exposition *f*
e instalación *f*
i esposizione *f*
p exposição *f*
d Aufstellung *f*

1002 exsiccation
f dessiccation *f*
e desecación *f*
i disseccamento *m*
p exsicação *f*
d Exsikkation *f*

1003 extensive *adj*
f étendu
e extenso
i estensivo
p extensivo
d ausgedehnt

1004 extensive pasture
f pâturage *m* extensif
e pastoreo *m* extensivo
i pascolo *m* estensivo
p pastagem *f* extensiva
d Extensivweide *f*

1005 extinction of water-vapour
f extinction *f* par la vapeur d'eau
e extinción *f* por el vapor acuoso
i estinzione *f* dovuta al vapore acqueo
p extinção *f* pelo vapor d'água
d Wasserdampfextinktion *f*

1006 extinct volcano
f volcan *m* éteint
e volcán *m* extinto
i vulcano *m* estinto
p vulcão *m* extinto
d erloschener Vulkan *m*

1007 extraordinary wave
f onde *f* extraordinaire
e onda *f* extraordinaria
i onda *f* straordinaria
p onda *f* extraordinária
d außerordentliche Welle *f*

1008 extraterrestrial light
f lumière *f* extraterrestre
e luz *f* extraterrestre
i luce *f* extraterrestre
p luz *f* extraterrestre
d extraterrestrisches Licht *n*

1009 extremes
f extrêmes *mpl*
e extremas *fpl*
i estremi *mpl*
p extremas *fpl*
d Extreme *npl*

* **eye of depression → 776**

1010 eye of wind
f lit *m* du vent
e filo *m* del viento
i occhio *m* del vento
p olho *m* do vento
d Windauge *n*

F

1011 F1 layer
(the lower part of the F layer at a virtual height of 300 km during the night and 200 km during the day)
f couche *f* F1
e capa *f* F1
i strato *m* F1
p camada *f* F1
d F1-Schicht *f*

1012 F2 layer
(the upper part of the F layer at a virtual height of 225 km)
f couche *f* F2
e capa *f* F2
i strato *m* F2
p camada *f* F2
d F2-Schicht *f*

1013 Fahrenheit scale
f échelle *f* Fahrenheit
e escala *f* Fahrenheit
i scala *f* Fahrenheit
p escala *f* Fahrenheit
d Fahrenheitskala *f*

1014 Fahrenheit temperature
f température *f* Fahrenheit
e temperatura *f* Fahrenheit
i temperatura *f* Fahrenheit
p temperatura *f* Fahrenheit
d Fahrenheittemperatur *f*

1015 Fahrenheit thermometer
f thermomètre *m* de Fahrenheit
e termómetro *m* de Fahrenheit
i termometro *m* di Fahrenheit
p termômetro *m* de Fahrenheit
d Fahrenheitthermometer *n*

1016 fair *adj*; **partly cloudy** *adj*
f peu nuageux
e poco nubloso
i poco nuvoloso
p parcialmente nublado
d halbbedeckt

1017 fair weather aeroplane
f avion *m* pour temps calme
e avión *m* para buen tiempo
i aeroplano *m* per bel tempo
p avião *m* para tempo calmo
d Schönwetterflugzeug *n*

1018 fair weather cumulus
f cumulus *m* de beau temps
e cúmulo *m* de buen tiempo
i cumulo *m* di buono tempo
p cúmulo *m* de bom tempo
d Schönwetterkumulus *m*

* **fair wind → 2677**

1019 fall
f pente *f*; déclivité *f*
e pendiente *f*; salto *m*
i pendenza *f*; declivio *m*
p queda *f*; baixa *f*
d Gefälle *n*

* **fall → 738**

1020 fall *v* **calm**
f être pris par le calme
e caer en una calma
i essere colto dalla bonaccia
p cair em uma calma
d von Windstille befallen werden

1021 falling current of air; descending current of air
f courant *m* d'air descendant
e corriente *f* de aire descendente
i corrente *f* d'aria discendente
p corrente *f* de ar descendente
d abfallender Luftstrom *m*

1022 falling of the barometer
f abaissement *m* du baromètre
e bajada *f* del barómetro
i abbassamento *m* del barometro
p queda *f* de pressão barométrica
d Fall *m* des Barometers

* **falling tide → 898**

1023 falling time of the barometer
f inertie *f* du baromètre
e inercia *f* del barómetro
i tempo *m* di discesa del barometro
p inércia *f* do barômetro
d Fallzeit *f* des Barometers

* **fall of temperature → 739**

1024 fallstreaks
f queues *fpl* de pluie
e corrientes *fpl* de precipitación de una nube que no llegan a tierra

i correnti *fpl* di precipitazione che non giungono a terra
p traços *mpl* de chuva nas alturas
d Fallstreifen *mpl*

* **fall-wind → 1587**

1025 false cirrus
f faux cirrus *m*
e cirro *m* falso
i falso cirro *m*
p falso cirro *m*
d falscher Zirrus *m*

1026 family of depressions
f série *f* de dépressions
e serie *f* de depresiones
i serie *f* di depressioni
p série *f* de depressões
d Tiefdruckserie *f*

* **farming → 68**

1027 fast ice
f glace *f* fixe
e banco *m* costero de hielo
i ghiaccio *m* fisso
p gelo *m* firme da costa
d Festeis *n*

1028 fathom
f brasse *f*
e braza *f*
i braccio *m*
p braça *f*
d Lachter *n*

1029 favourable weather
f temps *m* favorable
e tiempo *m* favorable
i tempo *m* favorevole
p tempo *m* favorável
d günstiges Wetter *n*

1030 February
f février *m*
e febrero *m*
i febbraio *m*
p fevereiro *m*
d Februar *m*

1031 fertile soil
f sol *m* fertile
e suelo *m* fértil
i terreno *m* fertile
p solo *m* fértil
d fruchtbarer Boden *m*

* **field crop → 1983**

1032 field curing period
f période *f* de séchage en champ
e período *m* de secado en el campo
i periodo *m* di essiccazione in pieno campo
p período *m* de secagem no campo
d Feldtrocknungszeit *f*

1033 field heat
f chaleur *f* de récolte
e calor *m* de recolección
i calore *m* di campo
p calor *m* de campo
d Feldwärme *f*

1034 field resistance
f résistance *f* sur le champ
e resistencia *f* del campo
i resistenza *f* di campo
p resistência *f* do campo
d Feldresistenz *f*

1035 field retting
f rouissage *m* à la rosée
e enriamiento *m* en rocío
i macerazione *f* a rugiada
p maceração *f* em orvalho
d Tauröste *f*

1036 field vegetable farming
f culture *f* légumière de plein champ
e cultivo *m* de hortalizas en el campo
i orticoltura *f* di pieno campo
p cultura *f* de verduras no campo
d Feldgemüsebau *m*

1037 figure of the earth; shape of the earth
f figure *f* de la terre
e formato *m* de la tierra
i figura *f* della terra
p formato *m* da terra
d Erdfigur *f*; Gestalt *f* der Erde

1038 filling up of a depression
f comblement *m* d'une dépression
e henchimiento *m* de una depresión
i colmarsi *m* di una depressione
p enchimento *m* de uma depressão; enchimento *m* de um ciclone
d Auffüllen *n* einer Zyklone

1039 filtering; filtration
f filtrage *m*; filtration *f*
e filtración *f*
i filtratura *f*
p filtração *f*
d Filtern *n*

* **filtration → 1039**

* **fine weather → 2357**

1040 fire avalanche
f avalanche *f* de blocs incandescents
e alud *m* de blocos incandescentes
i valanga *f* di detriti ardenti
p avalancha *f* de blocos incandescentes
d glühender Detritusbrei *m*

1041 fireball
f boule *f* de feu
e bola *f* de fuego
i pallone *m* di fuoco
p bola *f* de fogo
d Feuerkugel *f*

1042 fire storm
f tempête *f* de feu
e tempestad *f* de fuego
i tempesta *f* di fuoco
p tempestade *f* de fogo
d Feuersturm *m*

* **firmament → 483**

1043 firn basin; firn field
f champ *m* de névé; névé *m*
e campo *m* de nieve
i campo *m* di firn; nevato *m*
p campo *m* de neve; nevada *f*
d Firnbecken *n*; Firn *m*

* **firn field → 1043**

1044 firnification of snow
f névéfication *f* de la neige
e formación *f* de nevada
i firnificazione *f* della neve
p formação *f* de neve granulada
d Schneeverfirnen *n*

1045 first meridian
f premier méridien *m*
e primero meridiano *m*
i primo meridiano *m*
p primeiro meridiano *m*
d erster Meridian *m*

1046 first quarter
f premier quartier *m*
e cuarto *m* cresciente
i primo quarto *m*
p quarto *m* crescente
d erstes Viertel *n*

1047 first-year ice; white ice
f glace *f* de première année
e hielo *m* del primer año
i ghiaccio *m* del primo anno
p gelo *m* do primeiro ano
d Wintereis *n*; einjähriges Meereis *n*

1048 five-crop farming
f assolement *m* quinquennal; rotation *f* quinquennale des cultures
e rotación *f* quinquenal
i rotazione *f* quinquennale
p rotação *f* quinquenal
d Fünffelderwirtschaft *f*

* **flake of snow → 2422**

1049 flaming aurora
f aurore *f* flamboyante
e aurora *f* flamante
i aurora *f* flammeggiante
p aurora *f* flamejante
d flammendes Polarlicht *n*

1050 F layer
(upper ionized layer of the ionosphere also called F region which consists of the F1 and F2 layers)
f couche *f* F
e capa *f* F
i strato *m* F
p camada *f* F
d F-Schicht *f*

1051 floating ice
f glace *f* flottante
e hielo *m* flotante
i ghiaccio *m* galleggiante
p gelo *m* flutuante
d Treibeis *n*

1052 floccus
f nuage *m* floconneux
e floco *m*
i fiocco *m*
p floco *m*
d Flokkuswolke *f*

1053 floeberg *m*
(large mass of floating sea ice)
f floeberg *m*
e floeberg *m*
i floeberg *m*
p floebergue *m*
d Scholleneisberg *n*

1054 flood *v*
f inonder
e inundar
i inondare
p inundar
d überschwemmen

1055 flood channel
f chenal *m* de flot
e canal *m* de desagüe

i canale *m* di piena
p canal *m* de desagüe de inundação
d Abflußkanal *m*

1056 flood control; flood protection
f lutte *f* contre les inondations; protection *f* contre les inondations
e protección *f* contra las inundaciones
i difesa *f* contro le inondazioni
p defesa *f* contra cheias; proteção *f* contra as inundações
d Hochwasserschutz *m*; Schutz *m* gegen Überschwemmungen

1057 flood control dam
f digue *f* de protection contre les inondations
e dique *m* de protección contra las inundaciones
i argine *m* di protezione contro inondazioni
p dique *m* de proteção contra cheias
d Hochwasserschutzdamm *m*

1058 flood dam
f digue *f* contre les crues
e dique *m* contra inundaciones
i argine *m* di piena
p dique *m* contra enchentes
d Hochwasserdamm *m*

1059 flooded region
f région *f* inondée
e región *f* inundada
i regione *f* inondata
p região *f* inundada
d überschwemmtes Gebiet *n*

1060 flood frequency
f fréquence *f* des crues
e frecuencia *f* de inundaciones
i frequenza *f* delle piene
p frequência *f* de cheias
d Hochwasserhäufigkeit *f*

* **flooding → 1523**

1061 flooding of the graft union
f noyade *f* de la greffe
e inundación *f* de injerto
i inondazione *f* dell'innesto
p afogamento *m* do enxerto
d im Saft Ersticken *n* des Reises

1062 flood lighting
f éclairage *m* à grands flots
e alumbrado *m* de grandes olas
i illuminazione *f* a grandi onde
p iluminação *f* de grandes ondas
d Flutbeleuchtung *f*

1063 flood opening
f ouverture *f* de décharge des crues
e abertura *f* de descarga de las crecidas
i abertura *f* di scarico delle piene
p abertura *f* de descarga das cheias
d Hochwasserdurchlaß *m*

1064 flood plain
f plaine *f* alluviale; lit *m* d'inondation
e zona *f* de inundación; llanura *f* de inundación
i piana *f* alluviale; piana *f* golenale
p zona *f* de inundação
d Hochflutebene *f*; Vorland *n*

1065 flood plain deposit
f dépôt *m* alluvial de crue
e depósito *m* aluvial de crecidas
i deposito *m* alluvionale di piena
p aluvião *m* de cheias
d Hochwasserablagerung *f*

* **flood protection → 1056**

1066 flood records
f statistique *f* des crues
e estadística *f* de inundaciones
i registro *m* delle piene
p estatística *f* de inundações
d Hochwasserstatistik *f*

1067 flood routing
f étude *f* de la progression des crues
e regulación *f* de las crecidas
i studio *m* della progressione delle piene
p propagação *f* de cheia
d Studium *n* der Flutwelle

1068 flood stage
f niveau *m* critique de crue
e altura *f* hidrométrica crítica
i livello *m* critico di piena
p nível *m* crítico da inundação
d kritischer Hochwasserstand *m*

* **floor heating → 2452**

1069 flower formation
f formation *f* de fleurs
e formación *f* de flores
i formazione *f* fiorale
p formação *f* de flores
d Blütenbildung *f*

1070 flowering; blossoming
f floraison *f*
e floración *f*
i fioritura *f*
p floração *f*
d Blüte *f*

1071 flowering period
f époque *f* de floraison
e período *m* de floración
i periodo *m* di fioritura
p período *m* de floração
d Blütezeit *f*

1072 flower initiation
f initiation *f* florale
e iniciación *f* floral
i iniziazione *f* fiorale
p iniciação *f* floral
d Blütenanlage *f*

1073 flower of compass points on the chart
f rose *f* de la carte
e rosa *f* de los vientos de la carta
i rosa *f* della carta
p rosa *f* dos ventos da carta
d Kartenrose *f*

1074 flow-off
f débit *m*
e caudal *m*
i deflusso *m*
p área *f* de inundação
d Abfluß *m*

1075 fluctuation of climate
f variation *f* du climat
e fluctuación *f* del clima
i variazione *f* del clima
p variação *f* do clima
d Klimaschwankung *f*

1076 fluorescence
f fluorescence *f*
e fluorescencia *f*
i fluorescenza *f*
p fluorescência *f*
d Fluoreszenz *f*

1077 fluvial deposit
f sédiment *m* fluviatile
e sedimento *m* fluvial
i deposito *m* fluviale
p depósito *m* fluvial
d Flußablagerung *f*

1078 fluvioglacial *adj*
f fluvioglaciaire
e fluvioglacial
i fluvioglaciale
p fluvioglacial
d fluvioglazial

1079 foam
f écume *f*
e espuma *f*
i schiuma *f*
p espuma *f*
d Schaum *m*

1080 focal depth
f profondeur *f* du foyer
e focal *m* hipocentral
i profondità *f* ipocentrale
p profundidade *f* hipocentral
d Herdtiefe *f*

1081 focal height
f hauteur *f* focale
e altura *f* focal
i altezza *f* focale
p altura *f* focal
d Brennhöhe *f*

1082 foehn
(warm, dry wind, prevalent on the northern slopes of the Alps)
f foehn *m*
e foehn *m*
i foehn *m*
p foehn *m*
d Föhn *m*

1083 fog
f brouillard *m*
e niebla *f*
i nebbia *f*
p bruma *f*; nevoeiro *m*
d Nebel *m*

1084 fog bank
f banc *m* de brume
e banco *m* de calina
i banco *m* di nebbia
p banco *m* de nevoeiro
d Nebelbank *f*

1085 fog bell
f cloche *f* de brouillard
e campaña *f* de bruma
i campana *f* da nebbia
p sino *m* anunciador de nevoeiro
d Nebelglocke *f*

* **fogbow → 3004**

* **fogdog → 3004**

* **fog drip → 547**

1086 fog gong
f gong *m* de brume
e gong *m* de niebla
i gong *m* da nebbia
p gongo *m* anunciador de nevoeiro
d Nebelgong *m*

1087 fog gun
f canon *m* de brume
e cañón *m* de niebla
i cannone *m* da nebbia
p canhão *m* anunciador de nevoeiro
d Nebelknallsignal *n*

* **foggy** *adj* → **1320**

1088 foggy weather; hazy weather
f temps *m* brumeux
e tiempo *m* calinoso
i tempo *m* caliginoso
p tempo *m* brumoso; tempo *m* nevoento
d nebliges Wetter *n*

* **fog, in a ~ → 1487**

1089 fog lamp
f phare *f* antibrouillard
e faro *m* antiniebla
i faro *m* di nebbia
p farol *m* de nevoeiro
d Nebellampe *f*

1090 fog-signalling plant
f installation *f* à signaux de brume
e instalación *f* de señales de niebla
i impianto *m* da segnali di nebbia
p instalação *f* de sinalização de nevoeiro
d Nebelsignalanlage *f*

1091 fog warning
f avertissement *m* de brume
e aviso *m* de niebla
i avviso *m* di nebbia
p aviso *m* de nevoeiro
d Nebelwarnung *f*

1092 fog whistle
f sifflet *m* de brume
e silbato *m* de niebla
i fischio *m* da nebbia
p apito *m* de nevoeiro
d Nebelpfeife *f*

1093 foliage
f feuillage *m*
e follaje *m*
i fogliame *m*
p folhagem *f*
d Laub *n*

1094 foliation
f feuillaison *f*
e foliación *f*
i foliazione *f*
p folheação *f*
d Blattentwicklung *f*

1095 foot
(unit of length equal to 0.3048 m)
f pied *m*
e pie *m*
i piede *m*
p pé *m*
d Fuß *m*

1096 forbidden region; forbidden space
f région *f* interdite
e región *f* prohibida
i spazio *m* interdetto
p região *f* interditada
d verbotener Raum *m*

* **forbidden space → 1096**

1097 forced air circulation
f circulation *f* d'air forcée
e ventilación *f* de aire
i circolazione *f* forzata dell'aria
p ventilação *f* forçada do ar
d erzwungener Luftumlauf *m*

1098 forced ventilation
f ventilation *f* forcée
e ventilación *f* forzada
i ventilazione *f* forzata
p ventilação *f* forçada
d Zwangslüftung *f*

1099 forcing house
f serre *f* à forcer
e invernadero *m* de forzado
i serra *f* di forzatura
p invernadeiro *m* forçado
d Treibhaus *n*

1100 forecast; prediction
f prévision *f*
e pronóstico *m*
i previsione *f*
p previsão *f*
d Voraussage *f*

1101 forecasting map
f carte *f* prévisionnelle
e mapa *m* provisorio
i carta *f* previsionale
p mapa *m* de previsão
d Höffigkeitskarte *f*

* **forenoon → 1874**

1102 forenoon watch
f quart *m* de 8 h à midi
e cuarto *m* de la mañana
i guardia *f* del mattino
p quarto *m* de manhã
d Vormittagswache *f (von 8 bis 12 Uhr)*

1103 forest
f forêt *f*
e bosque *m*
i foresta *f*
p floresta *f*
d Wald *m*

1104 forest humus
f humus *m* forestier
e humus *m* forestal
i humus *m* forestale
p húmus *m* florestal
d Waldhumus *m*

1105 forest litter
f litière *f* forestière
e hojarasca *f* forestal
i lettiera *f* di foresta
p manta *f* morta da floresta
d Waldstreu *f*

1106 forest planting stock
f matériel *m* de plantation forestière
e material *m* de plantación forestal
i materiale *m* de piantagione forestale
p material *m* florestal para plantação
d Pflanzgut *n* für Wald und Forst

1107 forest reserves
f réserves *fpl* forestières
e reservas *fpl* forestales
i riserve *fpl* forestali
p reservas *fpl* florestais
d Waldreserven *fpl*

1108 forest soil
f terre *f* de forêt; sol *m* forestier
e suelo *m* de bosque; suelo *m* forestal
i terreno *m* forestale
p solo *m* florestal
d Waldboden *m*

1109 formation of clouds
f formation *f* de nuages
e formación *f* de nubes
i formazione *f* di nubi
p formação *f* de nuvens
d Wolkenbildung *f*

1110 formation pruning
f taille *f* de formation
e poda *f* de formación
i potatura *f* di formazione
p poda *f* de formação
d Erziehungsschnitt *m*

1111 forms without ray structure
f formes *fpl* sans structure de rayons
e formas *fpl* sin estructura de rayos
i forme *fpl* senza struttura di raggi
p formas *fpl* sem estrutura de raios
d Formen *fpl* ohne Strahlenstruktur

1112 forms with ray structure
f formes *fpl* avec structure de rayons
e formas *fpl* con estructura de rayos
i forme *fpl* con struttura a raggi
p formas *fpl* com estrutura de raios
d Formen *fpl* mit Strahlenstruktur

1113 Fortin's barometer
f baromètre *m* de Fortin
e barómetro *m* de Fortin
i barometro *m* di Fortin
p barômetro *m* de Fortin
d Fortinsches Barometer *n*

1114 fossil ice
f glace *f* fossile
e hielo *m* fósil
i ghiaccio *m* fossile
p gelo *m* fóssil
d fossiles Eis *n*

1115 foul air; polluted air
f air *m* vicié
e aire *m* viciado
i aria *f* impura
p ar *m* impuro
d unreine Luft *f*

1116 foul weather
f sale temps *m*
e tiempo *m* sucio
i tempo *m* sporco
p mau tempo *m*
d schlechtes Wetter *n*

1117 fractocumulus
(type of cumulus cloud characterized by its small size and often rapidly changing)
f fracto-cumulus *m*
e fractocumulus *m*
i fractocumulo *m*
p fractocumulus *m*
d Fraktokumulus *m*

1118 fractostratus
(type of stratus cloud characterized by its small size and often rapidly changing)
f fracto-stratus *m*
e fractostratus *m*
i fractostrato *m*
p fractostratus *m*
d Fraktostratus *m*

1119 frame
f châssis *m*
e cajonera *f*
i cassone *m*
p chassi *m*
d Kasten *m*

1120 frazil ice
f frazil *m*
e nieve *f* fluvial flotante
i ghiaccio *m* in aghetti
p gelo *m* fluvial flutuante
d freischwebende Eiskristalle *npl* und Grundeis *n*

1121 free air
f air *m* libre
e aire *m* libre
i aria *f* libera
p ar *m* livre
d Freiluft *f*

1122 free-air reduction
f réduction *f* à l'air libre
e reducción *f* de aire libre
i riduzione *f* all'aria libera
p diminuição *f* ao ar-livre
d Freiluftreduktion *f*

1123 freeze-up
f prise *f* de la glace
e formación *f* de hielo
i congelamento *m* in superficie
p congelação *f* contínua
d Eisstand *m*

1124 freezing
f congélation *f*
e congelación *f*
i congelamento *m*
p congelamento *m*
d Gefrieren *n*

1125 freezing point; ice point
f point *m* de congélation
e punto *m* de congelación
i punto *m* di congelamento
p ponto *m* de congelação
d Gefrierpunkt *m*

1126 freezing rain
f pluie *f* verglacante
e lluvia *f* engelante
i pioggia *f* che gela
p chuva *f* glacial
d Hagelregen *m*

1127 freezing zone
f zone *f* de gel
e zona *f* afectada por la helada
i zona *f* di congelamento
p zona *f* de congelação
d Gefrierzone *f*

1128 frequency
f fréquence *f*
e frecuencia *f*
i frequenza *f*
p frequência *f*
d Frequenz *f*

1129 frequency distribution
f distribution *f* de fréquence
e distribución *f* de la frecuencia
i distribuzione *f* di frequenza
p distribuição *f* de frequência
d Frequenzverteilung *f*

1130 fresh air
f air *m* frais
e aire *m* fresco
i aria *f* fresca
p ar *m* fresco
d frische Luft *f*

1131 fresh breeze
f bonne brise *f*
e viento *m* leve
i vento *m* teso
p brisa *f* fresca
d frische Brise *f*

1132 freshening wind
f vent *m* fraîchissant
e viento *m* refrescador
i vento *m* rinfrescante
p vento *m* refrescador
d auffrischender Wind *m*

*** fresh gale → 2591**

1133 fresh mass
f masse *f* verte
e masa *f* fresca
i massa *f* verde
p massa *f* fresca
d Frischmasse *f*

1134 friction
f frottement *m*
e rozamiento *m*
i frizione *f*
p fricção *f*
d Reibung *f*

1135 friction layer
f couche *f* de friction
e capa *f* de fricción
i strato *m* d'attrito
p camada *f* de atrito
d Reibungsschicht *f*

1136 frigid zone
f zone *f* glaciale
e zona *f* glacial
i zona *f* glaciale
p zona *f* glacial
d kalte Zone *f*

* **fringe region → 997**

1137 front
(the boundary between two air masses of different density or temperature)
f front *m*
e frente *m*
i fronte *m*
p frente *f*
d Front *f*

1138 frontal thunderstorm
f orage *m* frontal
e turbonada *f* frontal
i tempesta *f* frontale
p trovoada *f* frontal
d frontale Gewitterbö *f*

1139 frontogenesis
f frontogenèse *f*
e frontogénesis *f*
i frontogenesi *f*
p frontogênese *f*
d Frontogenese *f*

1140 frontolysis
f frontolyse *f*
e frontólisis *f*
i frontolisi *f*
p frontólise *f*
d Frontolyse *f*

1141 frost *v*
f geler
e helar
i gelare
p gear
d gefrieren

1142 frost action
f action *f* de froid
e acción *f* del frío
i azione *f* del gelo
p ação *f* de congelamento
d Kältewirkung *f*

1143 frost boil
f éminence *f* due au gel
e ampolla *f* de congelación
i bolla *f* dovuta al gelo
p empolamento *m* causado pela geada
d Frostbeule *f*

1144 frost crack
f fente *f* de gel
e grieta *f* de helada
i crepa *f* danno da gelo
p fissura *f* provocada pela geada
d Frostriß *m*

1145 frost damage; frost injury
f dégât *m* par la gelée
e daño *m* causado por las heladas
i danno *m* da gelo
p dano *m* causado pela geada
d Frostschaden *m*

1146 frost day
f jour *m* de gel
e día *f* de helada
i giorno *m* de gelo
p dia *m* de geada
d Frosttag *m*

1147 frost hardiness
f résistance *f* au gel
e resistencia *f* a las heladas
i resistenza *f* al gelo
p resistência *f* às geadas
d Frosthärte *f*; Frostresistenz *f*

1148 frost heaving
f gonflement *m* au gel
e deformación *f* del hielo
i deformazione *f* da gelo
p deformação *f* do gelo
d Frosthebung *f*

* **frost injury → 1145**

1149 frostless zone
f zone *f* verdoyante
e zona *f* sin helada
i zona *f* priva di gelate
p zona *f* sem geada
d Grüngürtel *m*

1150 frost lifting
f déchaussement *m* par la gelée
e descalzamiento *m* por heladas
i sradicamento *m* causato del gelo
p descalçamento *m* pela geada
d Auffrieren *n*

1151 frost penetration
f pénétration *f* du gel
e penetración *f* de la helada
i penetrazione *f* del gelo
p penetração *f* da geada
d Frosteindringung *f*

1152 frost pocket *(in fruit)*
f poche *f* de froid
e bolsa *f* de heladas
i sacca *f* di gelo
p bolsa *f* de geada
d Frosttasche *f*

1153 frost precaution
f précaution *f* contre le gel
e precaución *f* contra heladas
i precauzione *f* contro il gelo
p proteção *f* contra o frio
d Frostschutzvorkehrung *f*

1154 frostproof *adj*
f résistant à la gelée
e incongelable
i resistente al gelo
p resistente à geada
d frostbeständig

1155 frost protection; protection against frost
f protection *f* antigel; protection *f* antigelée
e protección *f* contra las heladas
i difesa *f* antigelo
p proteção *f* contra geadas
d Frostschutz *m*

1156 frost protection equipment
f appareils *mpl* de protection contre les gelées
e aparatos *mpl* de protección contra las heladas
i apparecchi *mpl* per la lotta antigelo
p equipamentos *mpl* de proteção contra as geadas
d Frostschutzgerät *n*

* **frost smoke → 2330**

1157 frost-susceptible *adj*
f sensible au froid
e sensible al frío
i sensibile al gelo
p sensível ao frio
d frostempfindlich

1158 frosty *adj*
f glacial
e helado; gélido
i gelido
p gelado
d frostig

1159 fructification
f fructification *f*
e fructificación *f*
i fruttificazione *f*
p frutificação *f*
d Fruchtbildung *f*

1160 fruit plantation
f plantation *f* fruitière
e plantación *f* frutal
i piantagione *f* di alberi da frutto
p plantação *f* de árvores de fruto
d Obstanlage *f*

1161 fruit setting
f mise *f* à fruit
e cuajado *m* de fruto
i allegazione *f*
p vingamento *m* do fruto
d Fruchtansatz *m*

1162 fulguration
f fulguration *f*
e fulguración *f*
i folgorazione *f*
p fulguração *f*
d Fulguration *f*

1163 full-blossom spraying
f pulvérisation *f* en pleine floraison
e pulverización *f* en plena floración
i trattamento *m* in piena fioritura
p pulverização *f* em plena floração
d Vollblütespritzung *f*

1164 full moon
f pleine lune *f*
e plenilunio *m*
i plenilunio *m*
p lua *f* cheia
d Vollmond *m*

1165 funnel cloud
f entonnoir *m* de la trombe
e nube *f* en embudo
i nube *f* a imbuto
p nuvem *f* em funil
d Trichter *m*; Schlauch *m*; Rüsselwolke *f*

G

1166 galactic light
f lumière *f* galactique
e luz *f* galáctica
i luce *f* galattica
p luz *f* galáctica
d galaktisches Licht *n*

1167 galaxy
f galaxie *f*
e galaxia *f*
i galassia *f*
p galáxia *f*
d Milchstraße *f*

1168 gale force
f force *f* de tempête
e fuerza *f* de tempestad
i forza *f* di temporale
p força *f* de tempestade
d Sturmkraft *f*

1169 gale warning; storm warning
f avis *m* de tempête; avertissement *m* de tempête
e aviso *m* de tempestad; aviso *m* de tormentas
i avviso *m* di temporale; segnale *m* di tempesta
p aviso *m* de tempestade; sinal *m* de temporal
d Sturmwarnung *f*; Sturmsignal *n*

1170 gallon
f gallon *m*
e galón *m*
i gallone *m*
p galão *m*
d Gallone *f*

1171 gas
f gaz *m*
e gas *m*
i gas *m*
p gás *m*
d Gas *n*

1172 Gay-Lussac's barometer
f baromètre *m* de Gay-Lussac
e barómetro *m* de Gay-Lussac
i barometro *m* di Gay-Lussac
p barômetro *m* de Gay-Lussac
d Gay-Lussacbarometer *n*

1173 gegenschein
f gegenschein *m*
e gegenschein *m*
i antichiarore *m*; anteliale *m*; gegenschein *m*
p luz *f* anti-solar; gegenschein *m*
d Gegenschein *m*

1174 general circulation
f circulation *f* générale
e circulación *f* general
i circolazione *f* generale
p circulação *f* geral
d allgemeine Zirkulation *f*

1175 general inference
f prévision *f* générale
e inferencia *f* general
i inferenza *f* generale
p inferência *f* geral
d allgemeine Wetterlage *f*

1176 gentle breeze; light wind; little wind
f petite brise *f*
e viento *m* flojo; brisa *f* débil
i brezza *f* tesa
p brisa *f* débil; vento *m* fraco
d schwache Brise *f*

1177 geocentric *adj*
f géocentrique
e geocéntrico
i geocentrico
p geocêntrico
d geozentrisch

1178 geodesy
f géodésie *f*
e geodesía *f*
i geodesia *f*
p geodésia *f*
d Geodäsie *f*

1179 geographic *adj*
f géographique
e geográfico
i geografico
p geográfico
d geographisch

1180 geographical equator
f équateur *m* géographique
e ecuador *m* geográfico
i equatore *m* geografico
p equador *m* geográfico
d geographischer Äquator *m*

1181 geographical meridian
f méridien *m* géographique
e meridiano *m* geográfico
i meridiano *m* geografico
p meridiano *m* geográfico
d geographischer Meridian *m*

1182 geographical pole
f pôle *m* géographique
e polo *m* geográfico
i polo *m* geografico
p pólo *m* geográfico
d geographischer Pol *m*

1183 geographic coordinates
f coordonnées *fpl* géographiques
e coordenadas *fpl* geográficas
i coordinate *fpl* geografiche
p coordenadas *fpl* geográficas
d geographische Koordinaten *fpl*

1184 geography
f géographie *f*
e geografía *f*
i geografia *f*
p geografia *f*
d Geographie *f*

1185 geoid
f géoïde *m*
e geoide *m*
i geoide *m*
p geóide *m*
d Geoid *n*

1186 geoisotherm
f isogéotherme *f*
e geoisoterma *f*
i geoisoterma *f*
p geoisoterma *f*
d Geoisotherme *f*

1187 geological *adj*
f géologique
e geológico
i geologico
p geológico
d geologisch

1188 geology
f géologie *f*
e geología *f*
i geologia *f*
p geologia *f*
d Geologie *f*

1189 geomagnetic *adj*
f géomagnétique
e geomagnético
i geomagnetico
p geomagnético
d geomagnetisch

1190 geomagnetic activity
f activité *f* géomagnétique
e actividad *f* geomagnética
i attività *f* geomagnetica
p atividade *f* geomagnética
d erdmagnetische Aktivität *f*

1191 geomagnetic elements
f éléments *mpl* géomagnétiques
e elementos *mpl* geomagnéticos
i elementi *mpl* geomagnetici
p elementos *mpl* geomagnéticos
d erdmagnetische Elemente *npl*

1192 geomagnetic storm
f orage *m* géomagnétique
e tempestad *f* geomagnética
i tempesta *f* geomagnetica
p tempestade *f* geomagnética
d erdmagnetischer Sturm *m*

1193 geometric *adj*
f géométrique
e geométrico
i geometrico
p geométrico
d geometrisch

1194 geometry
f géométrie *f*
e geometría *f*
i geometria *f*
p geometria *f*
d Geometrie *f*

1195 geophysics
f géophysique *f*
e geofísica *f*
i geofisica *f*
p geofísica *f*
d Geophysik *f*

1196 geopotential height
f hauteur *f* géopotentielle
e altura *f* geopotencial
i altitudine *f* geopotenziale
p altura *f* geopotencial
d geopotentielle Höhe *f*

1197 geostationary satellite
f satellite *m* géostationnaire
e satélite *m* geoestacionario
i satellite *m* geostazionario
p satélite *m* geoestacionário
d geostationärer Satellit *m*

1198 geostrophic wind
f vent *m* géostrophique
e viento *m* geostrófico
i vento *m* geostrofico
p vento *m* geostrófico
d geostrophischer Wind *m*

1199 geostrophic wind speed
f vitesse *f* du vent géostrophique
e velocidad *f* del viento geostrófico
i velocità *f* del vento geostrofico
p velocidade *f* do vento geostrófico
d geostrophische Windgeschwindigkeit *f*

1200 geosyncline
f géosynclinal *m*
e geosinclinal *m*
i geosinclinale *f*
p geossinclinal *m*
d Geosynklinale *f*

1201 geotectonics
f géotectonique *f*
e geotectónica *f*
i geotettonica *f*
p geotectônica *f*
d Geotektonik *f*

1202 geothermal gradient
f gradient *m* géothermique
e gradiente *m* geotérmico
i gradiente *m* geotermico
p gradiente *m* geotérmico
d geothermischer Gradient *m*

1203 geothermic *adj*
f géothermique
e geotérmico
i geotermico
p geotérmico
d geothermisch

1204 geothermic step
f degré *m* géothermique
e escalón *m* geotérmico
i grado *m* geotermico
p escala *f* geotérmica
d geothermische Tiefenstufe *f*

1205 geothermometer; earth thermometer
f géothermomètre *m*; thermomètre *m* dans le sol
e geotermómetro *m*; termómetro *m* de suelo
i geotermometro *m*
p geotermômetro *m*
d Geothermometer *n*

1206 geotropism
f géotropisme *m*
e geotropismo *m*
i geotropismo *m*
p geotropismo *m*
d Geotropismus *m*

1207 germination
f germination *f*
e germinación *f*
i germinazione *f*
p germinação *f*
d Keimung *f*

1208 germination period
f période *f* de germination
e período *m* de germinación
i periodo *m* di germinazione
p período *m* de germinação
d Keimzeit *f*

1209 geyser
f geyser *m*
e géiser *m*
i geyser *m*
p gêiser *m*
d Geyser *n*

1210 glacial *adj*
f glaciaire
e glacial
i glaciale
p glacial
d glazial

1211 glacial anticyclone
f anticyclone *m* glaciaire
e anticiclón *m* glacial
i anticiclone *m* glaciale
p anticiclone *m* glacial
d glazialer Antizyklon *m*

1212 glacial climate
f climat *m* glaciaire
e clima *m* glacial
i clima *m* glaciale
p clima *m* glacial
d vergletscherungsgünstiges Klima *n*

1213 glacial cycle
f cycle *m* glaciaire
e ciclo *m* glacial
i ciclo *m* glaciale
p ciclo *m* glacial
d glazialer Kreislauf *m*

1214 glacial deposits
f dépôts *mpl* glaciaires
e depósitos *mpl* glaciales

i depositi *mpl* glaciali
p depósitos *mpl* glaciais
d Gletschersedimentation *f*

1215 glacial environment; glacial medium
f milieu *m* glaciaire
e medio *m* glacial
i ambiente *m* glaciale
p ambiente *m* glacial
d Glazialmilieu *n*

1216 glacial erosion
f érosion *f* glaciaire
e umbral *m* glacial
i soglia *f*
p erosão *f* glacial
d Gletschererosion *f*

1217 glacial flood
f crue *f* glaciaire
e marea *f* glacial
i marea *f* glaciale
p maré *f* glacial
d glaziales Hochwasser *n*

1218 glacial lake
f lac *m* glaciaire
e lago *m* glacial
i lago *m* glaciale
p lago *m* glacial
d Gletschersee *f*

*** glacial medium → 1215**

1219 glacial valley
f vallée *f* glaciaire
e valle *m* glacial
i valle *m* glaciale
p vale *m* glacial
d Gletschertal *n*

*** glaciation → 1438**

1220 glaciation limit
f niveau *m* de glaciation
e límite *m* de glaciación
i limite *m* di glaciazione
p limite *m* de glaciação
d Vergletscherungsgrenze *f*

1221 glacier
f glacier *m*
e glaciar *m*
i ghiacciaio *m*
p glaciar *m*
d Gletscher *m*

1222 glacier activity
f activité *f* glaciaire
e actividad *f* glacial
i attività *f* glaciale
p atividade *f* glacial
d Gletscheraktivität *f*

1223 glacier breeze
f brise *f* de glacier
e brisa *f* de glaciar
i brezza *f* di ghiacciaio
p brisa *f* de glaciar
d Gletscherwind *m*

1224 glacier coefficient
f coefficient *m* glaciaire
e coeficiente *m* de glaciar
i coefficiente *m* di ghiacciaio
p coeficiente *m* de glaciar
d Gletscherkoeffizient *m*

1225 glacier desert
f désert *m* de glacier
e desierto *m* de glaciar
i deserto *m* di ghiacciaio
p deserto *m* de glaciar
d Eiswüste *f*

1226 glacier ice
f glace *f* de glacier
e hielo *m* de glaciar
i ghiaccio *m* di ghiacciaio
p gelo *m* de glaciar
d Gletschereis *n*

1227 glacier snout
f porte *f* de glacier
e boca *f* del glaciar
i bocca *f* di ghiacciaio
p boca *f* do glaciar
d Gletschertor *m*

1228 glacier table
f table *m* de glacier
e mesa *f* del glaciar
i blocco *m* di roccia in un ghiacciaio
p mesa *f* de glaciar
d Gletschertisch *m*

1229 glaciohydrology
f hydrologie *f* glaciaire
e hidrología *f* glacial
i idrologia *f* glaciale
p hidrologia *f* glacial
d Glazialhydrologie *f*

1230 glaciology
f glaciologie *f*
e glaciología *f*

i glaciologia *f*
p glaciologia *f*
d Glaziologie *f*

1231 glaciometeorology
f météorologie *f* glaciaire
e meteorología *f* glacial
i meteorologia *f* glaciale
p meteorologia *f* glacial
d Glazialmeteorologie *f*

*** glasshouse → 1262**

1232 glasshouse vegetables
f légumes *mpl* de serre
e hortaliza *f* de invernadero
i ortaggi *mpl* di serra
p legumes *mpl* de estufa
d Treibgemüse *n*

1233 glaze
f bruine *f* givrante
e cencellada *f* transparente
i ghiaccio *m* vetroso
p gelo *m* claro
d gefrierender Nieselregen *m*

1234 glaze *v*
f vitrer
e acristalar
i coprire da vetri
p envidraçar
d verglasen

*** glazed frost → 2206**

1235 gley soil; wet soil
f sol *m* à gley; sol *m* hydromorphe
e suelo *m* de gley; tierra *f* húmeda
i gley *m*; terreno *m* idromorfo
p solo *m* úmido
d Gleyboden *m*; Naßboden *m*

1236 gloomy weather
f temps *m* sombre
e tiempo *m* sombrio
i tempo *m* scuro
p tempo *m* sombrio
d trübes Wetter *n*

1237 glory
f gloire *f*
e aureola *f*
i aureola *f*
p auréola *f*
d Aureole *f*

1238 glory-of-snow
f chionodoxa *m*
e gloria *f* de la neve
i chionodoxa *f*
p glória *f* da neve
d Schneeglanz *m*

*** glow → 1689**

1239 glycerine barometer
f baromètre *m* à glycérine
e barómetro *m* de glicerina
i barometro *m* di glicerina
p barômetro *m* de glicerina
d Glyzerinbarometer *n*

*** GMT → 1269**

1240 gnomon
(column indicating the meridian altitude of the sun showing the time by its shadow)
f gnomon *m*
e gnomon *m*
i gnomone *m*
p gnômon *m*
d Gnomon *n*

1241 gnomonic projection
f projection *f* gnomonique
e proyección *f* gnomónica
i proiezione *f* gnomonica
p projeção *f* gnomônica
d gnomonische Projektion *f*

1242 good visibility
f bonne visibilité *f*
e buena visibilidad *f*
i buona visibilità *f*
p boa visibilidade *f*
d gute Sicht *f*

1243 gradient
f gradient *m*
e gradiente *m*
i gradiente *m*
p gradiente *m*
d Steigung *f*

1244 gradient wind
f vent *m* du gradient
e viento *m* del gradiente
i vento *m* di gradiente
p vento *m* de gradiente
d Gradientwind *m*

1245 gradient wind speed
f vitesse *f* gradientale du vent
e velocidad *f* gradiental del viento
i velocità *f* gradientale del vento

p velocidade *f* gradiental do vento
d Gradientwindgeschwindigkeit *f*

1246 gradual change of temperature
f variation *f* progressive de température
e variación *f* progresiva de temperatura
i variazione *f* progressiva di temperatura
p variação *f* progressiva de temperatura
d allmähliche Temperaturänderung *f*

1247 grain conditioning
f conditionnement *m* de grains
e acondicionamiento *m* de granos
i condizionamento *m* dei cereali
p acondicionamento *m* de cereais
d Getreideaufbereitung *f*

1248 grain moisture tester
f contrôleur *m* d'humidité de céréales
e indicador *m* de humedad de granos
i misuratore *m* dell'umidità delle granaglie
p indicador *m* de umidade de grãos de cereais
d Körnerfeuchtigkeitsmesser *m*

1249 gram
f gramme *m*
e gramo *m*
i grammo *m*
p grama *m*
d Gramm *n*

1250 granular *adj*
f granuleux; granulé
e granular
i granulare
p granular
d körnig

1251 granular ice
f glace *f* granulée; glace *f* granulaire
e hielo *m* granular
i ghiaccio *m* granulare
p gelo *m* granular
d Eisgranulat *n*; Eiskörner *n*

1252 granular snow
f neige *f* granulaire; neige *f* en grains
e nieve *f* granulada
i neve *f* granulare
p neve *f* granular
d körniger Schnee *m*

1253 grass temperature
f température *f* du sol gazonné
e temperatura *f* junto al suelo
i temperatura *f* dell'erba
p temperatura *f* da relva
d Temperatur *f* am Erdboden

* **graupel → 2436**

1254 gravimetry
f gravimétrie *f*
e gravimetría *f*
i gravimetria *f*
p gravimetria *f*
d Gravimetrie *f*

* **gravitational anomaly → 1256**

1255 gravity
f pesanteur *f*
e gravedad *f*
i gravità *f*
p gravidade *f*
d Schwerkraft *f*

1256 gravity anomaly; gravitational anomaly
f anomalie *f* de la force de pesanteur
e anomalía *f* de gravidad
i anomalia *f* della gravità
p anomalia *f* da gravidade
d Schwereanomalie *f*

* **gravity wind → 1587**

1257 grazing season
f saison *f* de pâturage
e período *m* de pastoreo; época *f* de pastoreo
i epoca *f* di pascolo
p período *m* de pastoreio
d Weidezeit *f*

1258 grease ice
f sorbet *m*
e hielo *m* grasoso
i ghiaccio *m* brillante
p gelo *m* gorduroso
d Eisschlamm *m*

1259 great circle
f grand cercle *m*
e círculo *m* máximo
i gran circolo *m*
p círculo *m* máximo
d Großkreis *m*

1260 greco
(northeast wind in Italy)
f greco *m*
e greco *m*
i greco *m*
p greco *m*
d Greco *m*

1261 green flash
f rayon *m* vert
e rayo *m* verde
i raggio *m* verde

p raio *m* verde
d grüner Strahl *m*

1262 greenhouse; glasshouse
f serre *f*
e invernadero *m*; estufa *f*
i serra *f*
p estufa *f* fria
d Gewächshaus *n*

1263 greenhouse climate
f climat *m* de serre
e clima *m* del invernadero
i clima *m* della serra
p clima *m* de estufa
d Gewächshausklima *n*

1264 greenhouse cultivation
f culture *f* sous verre
e cultivo *m* bajo cristal
i coltura *f* sotto serra in vetro
p cultura *f* em estufa
d Unterglaskultur *f*

1265 greenhouse effect
f effet *m* serre
e efecto *m* invernadero
i effetto *m* di serra
p efeito *m* estufa
d Gewächshauseffekt *m*

1266 greenhouse plant
f plante *f* de serre
e planta *f* de invernadero
i pianta *f* di serra
p planta *f* de estufa
d Gewächshauspflanze *f*

1267 greenhouse soil
f sol *m* de serre
e suelo *m* de invernadero
i terreno *m* di serra
p solo *m* de estufa
d Gewächshausboden *m*

1268 Greenwich apparent time
f temps *m* apparent de Greenwich
e hora *f* aparente de Greenwich
i tempo *m* apparente di Greenwich
p hora *f* aparente de Greenwich
d scheinbare Greenwichzeit *f*

1269 Greenwich mean time; GMT; universal time
f temps *m* universel; TU
e hora *f* media de Greenwich; hora *f* universal
i ora *f* media di Greenwich; tempo *m* universale
p hora *f* média de Greenwich; tempo *m* universal
d mittlere Greenwichzeit *f*; MGZ; Weltzeit *f*

1270 Greenwich meridian
f méridien *m* de Greenwich
e meridiano *m* de Greenwich
i meridiano *m* di Greenwich
p meridiano *m* de Greenwich
d Greenwichmeridian *m*

1271 Greenwich time
f temps *m* de Greenwich
e tiempo *m* de Greenwich
i tempo *m* di Greenwich
p tempo *m* de Greenwich
d Greenwichzeit *f*

1272 grippal *adj*
f grippal
e gripal
i grippiale
p gripal
d grippös

1273 grippe
f grippe *f*
e gripe *f*
i grippe *f*
p gripe *f*
d Grippe *f*

* **ground avalanche → 1613**

1274 ground disposal of effluent
f décharge *f* d'effluent dans la terre
e descarga *f* de efluente en la tierra
i scarico *m* d'effluente nella terra
p descarga *f* de efluente na terra
d Beseitigung *f* der aktiven Ausströmung in der Erde

1275 ground fog
f brume *f* au sol
e niebla *f* al ras del suelo
i nebbia *f* al suolo; nebbia *f* bassa
p nevoeiro *m* no solo; nevoeiro *m* rasteiro
d Bodennebel *m*

1276 ground frost; soil frost
(temperature ⩽ 30.4°F that is damaging to growing vegetation)
f gelée *f* blanche; gelée *f* du sol
e helada *f* superficial del suelo; escarcha *f*
i terreno *m* ghiacciato; gelata *f* al suolo
p geada *f*
d Bodenfrost *m*

1277 ground ice
f glace *f* du sol
e hielo *m* del suelo
i ghiaccio *m* del suolo
p gelo *m* do solo
d Bodeneis *n*

1278 ground swell
f houle *f* de fond
e mar *m* de fondo
i mare *m* di fondo
p mar *m* de fundo
d Grunddünung *f*

1279 groundwater; phreatic water
f eau *f* souterraine; eau *f* phréatique
e agua *f* subterránea; agua *f* freática
i acqua *f* sotterranea
p água *f* subterrânea; água *f* freática
d Grundwasser *n*

1280 groundwater depth indicator
f niveau *m* de la nappe phréatique
e nivel *m* del agua freática
i indicatore *m* di profondità della falda freatica
p nível *m* da água freática
d Grundwasserspiegel *m*

1281 ground wave
f onde *f* de sol
e onda *f* terrestre
i onda *f* al suolo; onda *f* alla superficie
p onda *f* terrestre
d Bodenwelle *f*

1282 ground wind
f vent *m* au sol
e viento *m* al suelo
i vento *m* al suolo
p vento *m* de superfície
d Bodenwind *m*

1283 group flashing light
f éclats *mpl* groupés
e grupo *m* de luces de relámpagos
i gruppo *m* de luci a lampi
p grupo *m* de luzes de relâmpagos
d Gruppenfeuer *n*

1284 group occulting light
f occultations *fpl* groupées
e grupo *m* de luces de destellos
i gruppo *m* de luci a splendori
p grupo *m* de luzes de eclipses
d Gruppenblinkfeuer *n*

1285 growing season; vegetation period
f période *f* de végétation
e período *m* vegetativo
i stagione *f* di crescita; periodo *m* vegetativo
p estação *f* de crescimento
d Vegetationsperiode *f*; Vegetationszeit *f*

1286 Gulf Coast
f côte *f* mexicaine
e costa *f* de Golfo mejicano
i costa *f* di Golfo messicano
p costa *f* de Golfo mexicano
d Küste *f* des Golfes von Mexiko

1287 Gulf Stream
f Courant *m* du Golfe
e Corriente *f* del Golfo
i Corrente *f* del Golfo
p Corrente *f* de Golfo
d Golfstrom *m*

1288 gulf weed; sargasso
f sargasse *f*
e sargazo *m*
i sargasso *m*
p sargaço *m*
d Golfkraut *n*; Sargassokraut *n*

1289 Gulf wind
f vent *m* du Golfe
e viento *m* de Golfo
i vento *m* di Golfo
p vento *m* de Golfo
d Golfwind *m*

1290 gully erosion
f érosion *f* en ravins
e erosión *f* en cárcavas
i burronamento *m*; erosione *f* a baratri
p erosão *f* em ravina
d Grabenerosion *f*

1291 gust gradient distance
f gradient *m* de rafale
e distancia *f* del gradiente de racha
i distanza *f* del gradiente di raffica
p distância *f* do gradiente de rajada
d Böentiefe *f*

1292 gustiness; bumpiness; turbulence; turbulent air
f intensité *f* de rafale; turbulence *f* de l'air
e turbulencia *f*
i regime *m* di raffiche; turbolenza *f* dell'aria
p condição *f* de rajadas de vento frequentes; turbulência *f*
d Böigkeit *f*; Turbulenz *f*

1293 gust of wind
f bourrasque *f*
e racha *f* de viento
i folata *f* di vento
p rajada *f* de vento
d Windstoß *m*

1294 guttation
f guttation *f*
e gutación *f*
i guttazione *f*
p gutação *f*
d Guttation *f*

1295 gutter
f rigole *f*
e gotera *f*
i doccia *f*
p goteira *f*
d Graben *n*

1296 gyroscope
f gyroscope *m*
e giroscopio *m*
i giroscopio *m*
p giroscópio *m*
d Gyroskop *n*; Kreisel *m*

1297 gyroscopic motion
f mouvement *m* gyroscopique
e movimiento *m* giroscópico
i movimento *m* giroscopico
p movimento *m* giroscópico
d Kreiselbewegung *f*

H

1298 haar
(wet sea fog from eastern Scotland and northeastern England)
f haar *m*
e haar *m*
i haar *m*
p haar *m*
d kalter Nebel *m*

1299 habitat
f habitat *m*
e habitat *m* ecológico
i ambiente *m* ecologico
p estação *f* ecológica
d Standort *m*

1300 hail
f grêle *f*
e granizo *m*
i grandine *f*
p granizo *m*
d Hagel *m*

1301 hail *v*
f grêler
e granizar
i grandinare
p granizar
d hageln

*** hail damage → 1303**

1302 hail defense
f protection *f* contre la grêle
e protección *f* contra el granizo
i difesa *f* antigrandine
p proteção *f* contra o granizo
d Hagelschutz *m*

1303 hail injury; hail damage
f dégât *m* causé par la grêle
e daño *m* causado por el granizo
i danno *m* da grandine
p dano *m* causado por granizo
d Hagelschaden *m*

1304 hail squall
f grain *m* grêleux
e granizada *f*
i piovasco *m* con grandine
p tormenta *f* de saraiva
d Hagelbö *f*

1305 hailstone
f grain *m* de grêle
e partícula *f* de granizo
i chicco *m* di grandine
p pedra *f* de granizo
d Hagelkorn *n*

1306 hair hygrometer
f hygromètre *m* à cheveux
e higrómetro *m* de cabello
i igrometro *m* capillare
p higrômetro *m* capilar
d Haarhygrometer *n*

*** half flood → 1308**

1307 half moon
f demi-lune *f*
e media luna *f*
i mezzaluna *f*
p meia-lua *f*
d Halbmond *m*

1308 half tide; half flood
f mi-marée *m*
e media marea *f*
i mezza marea *f*
p meia-maré *f*
d Halbgezeiten *fpl*

1309 halo
(circle of light around the sun or moon, caused by diffraction through ice crystals)
f halo *m*
e halo *m*
i alone *m*
p halo *m*
d Hof *m*

1310 halo formation
f formation *f* halo
e formación *f* halo
i alonatura *f*
p formação *f* de halo
d Lichthofbildung *f*

1311 halo of 22′
f halo *m* principal; halo *m* de 22′
e halo *m* de 22′
i alone *m* di 22′
p halo *m* de 22′
d kleiner Ring *m*; 22′-Ring *m*

1312 halo of 46′
f grand halo *m*; halo *m* de 46′
e halo *m* de 46′

i alone *m* di 46′; alone *m* grande
p halo *m* de 46′
d großer Ring *m*; 46′-Ring *m*

1313 halo phenomenon
f phénomène *m* de halo
e fenómeno *m* de halo
i fenomeno *m* di alone
p fenômeno *m* de halo
d Haloerscheinung *f*

1314 hanging glacier
f glacier *m* suspendu
e glaciar *m* suspendido
i ghiaccio *m* pendente
p glaciar *m* suspenso
d Hängegletscher *m*

1315 hardening
f durcissement *m*
e endurecimiento *m*
i indurimento *m*
p endurecimento *m*
d Verhärtung *f*

1316 harmattan
(periodical hot dry, dust-bearing easterly wind from northeast to east-northeast prevailing in the West of Africa)
f harmattan *m*
e harmattan *m*
i harmattan *m*
p harmatão *m*
d Harmattan *m*

1317 harvest
f récolte *f*
e cosecha *f*
i raccolta *f*
p colheita *f*
d Ernte *f*

1318 hazard
f risque *m*
e peligro *m*
i rischio *m*
p perigo *m*
d Gefahr *f*

1319 haze
f brume *f* sèche
e calina *f*
i caligine *f*
p caligem *f*
d Dunst *m*

1320 hazy *adj*; **foggy** *adj*; **misty** *adj*
f brumeux
e brumoso
i nebbioso
p brumoso
d neblig; diesig

* **hazy weather → 1088**

* **head wind → 827**

1321 health resort
f station *f* climatique
e estación *f* climática
i stazione *f* climatica
p estação *f* climática
d Luftkurort *m*

1322 health resort in high altitude
f station *f* d'altitude
e estación *f* de altura
i stazione *f* dell'altitudine
p estação *f* hidroterápica de altura
d Höhenkurort *m*

1323 heap clouds; cloud streets
f nuages *mpl* en monceaux
e nubes *fpl* cumuliformes
i nubi *fpl* rampanti
p castelos *mpl* de nuvens
d Haufenwolken *fpl*

1324 heat
f chaleur *f*
e calor *m*
i calore *m*
p calor *m*
d Wärme *f*

1325 heat balance
f bilan *m* thermique; bilan *m* calorifique
e balance *m* térmico
i bilancio *m* termico
p balanço *m* térmico
d Wärmebilanz *f*

1326 heat capacity of snow
f capacité *f* calorifique de la neige
e capacidad *f* calorífica de la nieve
i capacità *f* calorifica della neve
p capacidade *f* calorífica da neve
d Wärmekapazität *f* von Schnee

1327 heat conductivity of snow
f conductibilité *f* thermique de la neige
e conductividad *f* térmica de la nieve
i conducibilità *f* termica della neve
p condutividade *f* térmica da neve
d Wärmeleitfähigkeit *f* von Schnee

1328 heat damage; heat injury
f dégât *m* dû à la chaleur
e daño *m* por insolación
i danno *m* da calore
p dano *m* por insolação
d Hitzeschaden *m*

1329 heated crop
f culture *f* chauffée
e cultivo *m* con calefacción
i coltura *f* in ambiente riscaldato
p cultura *f* aquecida
d geheizte Kultur *f*

1330 heat equilibrium
f équilibre *m* thermique
e equilibrio *m* térmico
i equilibrio *m* termico
p equilíbrio *m* térmico
d Wärmegleichgewicht *n*

1331 heat index
f indice *m* de chaleur
e índice *m* de calor
i indice *m* di calore
p índice *m* de calor
d Wärmezahl *f*

1332 heating by air circulation
f chauffage *m* à circulation d'air
e calefacción *f* por circulación de aire
i riscaldamento *m* a circolazione d'aria
p aquecimento *m* por circulação de ar
d Umlaufluftheizung *f*

* **heat injury → 1328**

1333 heat insulation
f calorifugeage *m*
e aislamiento *m* calórico
i isolamento *m* calorico
p isolamento *m* refratário
d Wärmeisolierung *f*

1334 heat of condensation
f chaleur *f* de condensation
e calor *m* de condensación
i calore *m* di condensazione
p calor *m* de condensação
d Kondensationswärme *f*

1335 heat of evaporation
f chaleur *f* d'évaporation
e calor *m* de evaporación
i calore *m* di evaporizzazione
p calor *m* de evaporação
d Verdunstungswärme *f*

1336 heat of vaporization
f chaleur *f* de vaporisation
e calor *m* de vaporización
i calore *m* di vaporizzazione
p calor *m* de vaporização
d Verdampfungswärme *f*

1337 heat sensation
f sensation *f* de chaleur
e sensación *f* de calor
i sensazione *f* di calore
p sensação *f* de calor
d Hitzesensation *f*

1338 heat stroke
f coup *m* de chaleur
e golpe *m* de calor
i colpo *m* di calore
p golpe *m* de calor
d Hitzeschlag *m*

* **heat sum → 2695**

1339 heat thunderstorm
f orage *m* de chaleur
e tormenta *f* de calor
i tempesta *f* di calore
p tormenta *f* de calor
d Wärmegewitter *n*

1340 heat wave
f vague *f* de chaleur
e ola *f* de calor
i onda *f* di calore
p onda *f* de calor
d Hitzewelle *f*

1341 Heaviside layer
f couche *f* de Heaviside
e estrato *m* de Heaviside
i strato *m* di Heaviside
p camada *f* de Heaviside
d Heavisidegürtel *m*

1342 heavy rain; rain shower
f grosse pluie *f*; torrent *m* de pluie
e chaparrón *m*; aguacero *m*
i pioggia *f* dirotta; acquazione *f*
p chuva *f* forte; enxurrada *f*
d schwerer Regen *m*; Regenschauer *m*

1343 hedge
f haie *f* vive
e cerco *m* vivo
i siepe *f* viva
p cerca *f* viva
d Hecke *f*

1344 height
f hauteur *f*
e altura *f*
i altezza *f*
p altura *f*
d Erhebung *f*

1345 height diagram
f diagramme *m* des hauteurs
e diagrama *m* de las alturas
i diagramma *m* delle altezze
p diagrama *m* das alturas
d Höhendiagramm *n*

1346 height of a layer
f hauteur *f* d'une couche
e altura *f* del estrato
i altezza *f* dello strato
p altura *f* da camada
d Schichthöhe *f*

1347 height of rainfall; rainfall height
f hauteur *f* de pluie
e altura *f* de lluvia
i altezza *f* di pioggia
p altura *f* de chuva
d Regenhöhe *f*

1348 height of the airglow layer; airglow height
f altitude *f* d'émission de la lueur nocturne
e altura *f* del resplandor del aire luminoso
i altezza *f* dello strato luminoso
p altura *f* da luz celeste
d Höhe *f* der Schicht des Nachthimmelleuchtens

1349 height of the moon
f hauteur *f* de la lune
e altura *f* de la luna
i altezza *f* della luna
p altura *f* da lua
d Mondhöhe *f*

1350 height of the tide
f hauteur *f* de la marée
e altura *f* de la marea
i altezza *f* della marea
p altura *f* da maré
d Höhe *f* der Gezeiten

* **height of wave → 2957**

1351 height table
f table *m* des hauteurs
e tabla *f* de alturas
i tavola *f* delle altezze
p tabela *f* de alturas
d Höhentafel *f*

1352 heliograph; sunshine recorder
f héliographe *m*
e heliógrafo *m*
i eliografo *m*
p heliógrafo *m*
d Heliograph *m*; Sonnenscheinautograph *m*

1353 heliotherapy
f héliothérapie *f*
e helioterapia *f*
i elioterapia *f*
p hélioterapia *f*
d Heliotherapie *f*

1354 heliothermometer
f héliothermomètre *m*
e heliotermómetro *m*
i termometro *m* a elio
p heliotermômetro *m*
d Heliothermometer *n*

1355 heliotrope
f héliotrope *m*
e heliotropo *m*
i eliotropo *m*
p heliotrópio *m*
d Heliotrop *m*

1356 helium
(one of the rare gases; symbol He, atomic number 2)
f hélium *m*
e helio *m*
i elio *m*
p hélium *m*
d Helium *n*

1357 hemisphere
f hémisphère *m*
e hemisferio *m*
i emisfero *m*
p hemisfério *m*
d Hemisphäre *f*

1358 heterosphere
f hétérosphère *f*
e heterosfera *f*
i eterosfera *f*
p heterosfera *f*
d Heterosphäre *f*

1359 hibernation
f hibernation *f*
e hibernación *f*
i ibernazione *f*
p hibernação *f*
d Hibernation *f*

* **high atmosphere → 2866**

1360 high clouds
f nuages *mpl* supérieurs
e nubes *fpl* altas
i nubi *fpl* alte
p nuvens *fpl* altas
d hohe Wolken *fpl*

1361 high frequency
f haute fréquence *f*
e alta frecuencia *f*
i alta frequenza *f*
p alta frequência *f*
d hohe Frequenz *f*

* **high frequency waves → 2374**

1362 high mountain
f haute montagne *f*
e alta montaña *f*
i alta montagna *f*
p alta montanha *f*
d Hochgebirge *n*

1363 high pressure area; anticyclone
f zone *f* de haute pression; anticyclone *m*
e zona *f* de alta presión; anticiclón *m*
i zona *f* di alta pressione; anticiclone *m*
p zona *f* de alta pressão; anticiclone *m*
d Hochdruckgebiet *n*; Antizyklon *m*

1364 high water
f pleine mer *m*
e marea *f* alta
i alta marea *f*
p maré *f* alta
d Hochwasser *n*

1365 high-water level
f niveau *m* de la pleine mer
e nivel *m* de la pleamar
i livello *m* dell'alta marea
p nível *m* da preamar
d Hochwasserebene *f*

1366 high-water line
f limite *f* de la marée
e línea *f* de la marea
i linea *f* dell'acqua alta
p linha *f* de preamar
d Strandwall *m*

1367 hill fog; upslope fog
f cumulus-brouillard *m* de montée; brouillard *m* élevé
e niebla *f* en las laderas de una montaña; niebla *f* elevada
i nebbia *f* sulle colline; nebbia *f* di collina
p nevoeiro *m* de encosta; nevoeiro *m* de um morro
d Bergnebel *m*; Berghangnebel *m*

1368 hillside up-current
f ascendance *f* de pente
e viento *m* de ladera
i corrente *f* di versante
p ascendência *f* da ladeira
d Hangwind *m*

1369 hoarfrost
(white deposit of ice crystals formed after a cold night by the sublimation of water on objects)
f gelée *f* blanche; givre *m*
e escarcha *f*
i brinata *f*
p geada *f* branca
d Rauhreif *m*

1370 hoarfrost formation
f formation *f* du givre mou
e formación *f* de escarcha
i formazione *f* di brina
p formação *f* de sincelos brancos
d Rauhreifbildung *f*

* **hoarfrost point → 790**

1371 hodograph
f hodographe *m*
e hodógrafo *m*
i odografo *m*
p hodógrafo *m*
d Hodograph *m*

1372 holocene
f holocène *m*
e holoceno *m*
i olocene *m*
p holoceno *m*
d Holozän *n*

1373 homogeneous atmosphere
f atmosphère *f* homogène
e atmósfera *f* homogénea
i atmosfera *f* omogenea
p atmosfera *f* homogênea
d homogene Atmosphäre *f*

1374 homogeneous band
f bande *f* homogène
e banda *f* homogénea
i banda *f* omogenea
p banda *f* homogênea
d homogene Bande *f*

1375 homogeneous quiet arc
f arc *m* homogène; arc *m* homogène calme
e arco *m* homogéneo
i arco *m* omogeneo quiescente
p arco *m* homogêneo
d homogener ruhiger Bogen *m*

1376 homopause
f homopause *f*
e homopausa *f*
i omopausa *f*
p homopausa *f*
d Homopause *f*

1377 homosphere
f homosphère *f*
e homosfera *f*
i omosfera *f*
p homosfera *f*
d Homosphäre *f*

1378 horizon
f horizon *m*
e horizonte *m*
i orizzonte *m*
p horizonte *m*
d Horizont *m*

1379 horizontal intensity (of magnetic field)
f intensité *f* horizontale (de champ magnétique)
e intensidad *f* horizontal (del campo magnético)
i intensità *f* orizzontale (del campo magnetico)
p intensidade *f* horizontal (do campo magnético)
d Horizontalintensität *f* (des Magnetfeldes)

1380 horse latitudes
f calmes *mpl* subtropicaux
e calmas *fpl* del trópico
i latitudini *fpl* delle calme
p região *f* de calmarias tropicais
d Roßbreiten *fpl*

1381 hot-air bath
f bain *m* d'air chaud
e baño *m* de aire caliente
i bagno *m* di aria calda
p banho *m* de ar quente
d Heißluftbad *n*

1382 hot-air heated crop
f culture *f* chauffée à l'air chaud
e cultivo *m* con calefacción por aire caliente
i coltura *f* riscaldata con aria calda
p cultura *f* aquecida a ar quente
d warmluftgeheizte Kultur *f*

1383 hot-air heating
f chauffage *m* à l'air chaud
e calefacción *f* por aire caliente
i riscaldamento *m* ad aria calda
p aquecimento *m* por ar quente
d Warmluftheizung *f*

1384 hot-air treatment
f traitement *m* à l'air chaud
e tratamiento *m* con aire caliente
i trattamento *m* con aria calda
p tratamento *m* com ar quente
d Heißluftbehandlung *f*

1385 hot blast
f vent *m* chaud
e viento *m* caliente
i corrente *f* d'aria calda
p vento *m* quente
d Heißluft *f*

1386 hot climate
f climat *m* chaud
e clima *m* cálido
i clima *m* caldo
p clima *m* quente
d heißes Klima *n*

1387 hot weather
f temps *m* très chaud
e tiempo *m* caluroso
i tempo *m* molto caldo
p tempo *m* muito quente
d heißes Wetter *n*

1388 hot-wire anemometer
f anémomètre *m* à fil chaud
e anemómetro *m* térmico
i anemometro *m* a filo caldo
p anemômetro *m* térmico
d Hitzdrahtanemometer *n*

1389 hot-wire microphone
f microphone *m* à fil chaud
e micrófono *m* de filo caliente
i microfono *m* a filo caldo
p microfone *m* de fio quente
d Hitzdrahtmikrophon *n*

1390 house heating
f chauffage *m* des immeubles
e calefacción *f* de las casas
i riscaldamento *m* delle abitazioni
p aquecimento *m* das casas
d Wohnungheizung *f*

* **HUM → 1398**

1391 human ecology
f écologie *f* humaine
e ecología *f* humana
i ecologia *f* umana
p ecologia *f* humana
d Sozialökologie *f*

* **Humboldt Current → 2045**

1392 humic gley soil
f sol *m* de prairie
e terreno *m* de prados
i terreno *m* prativo
p terreno *m* de prados
d Wiesenboden *m*

*** humid air → 1852**

1393 humid environment
f milieu *m* humide
e medio *m* húmedo
i ambiente *m* umido
p ambiente *m* úmido
d feuchtes Milieu *n*

1394 humidification
f humectation *f*
e humidificación *f*
i umidificazione *f*
p umidificação *f*
d Befeuchtung *f*

1395 humidify *v*
f humidifier; mouiller
e humectar; humedecer
i inumidire; umidificare
p umedecer
d befeuchten; anfeuchten

1396 humidity; moisture; dampness
f humidité *f*
e humedad *f*
i umidità *f*
p umidade *f*
d Feuchtigkeit *f*

1397 humidity ratio
f rapport *m* d'humidité
e relación *f* de humedad
i titolo *m* di umidità
p coeficiente *m* de umidade
d Feuchtigkeitsverhältnis *n*

1398 humilis; HUM
f humilis *m*
e humilis *m*
i humilis *m*
p humilis *m*
d Humiliswolke *f*

1399 hummock
f hummock *m*
e morón *m*
i poggio *m*
p montículo *m*
d Hummock; Packeishügel *m*

1400 hummocky ice
f glace *f* moutonnée
e protuberancias *fpl* del hielo
i monticello *m* di ghiaccio
p protuberâncias *fpl* no campo de gelo
d Eishöcker *m*

1401 hurricane; cyclonic storm
(tropical cyclone characterized by winds of force 12 (73 miles/117 kilometers per hour) on the Beaufort scale)
f ouragan *m*
e huracán *m*
i uragano *m*
p furacão *m*
d Hurrikan *m*; Orkan *m*

1402 hydraulics
f hydraulique
e hidráulica *f*
i idraulica *f*
p hidráulica *f*
d Hydraulik *f*

1403 hydrocarbon
f hydrocarbure *m*
e hidrocarburo *m*
i idrocarburo *m*
p hidrocarbono *m*
d Kohlenwasserstoff *m*

1404 hydrocyclone
f hydrocyclone
e hidrociclón *m*
i idrociclone *m*
p hidrociclone *m*
d Hydrozyklon *m*

1405 hydrogel
f hydrogel *m*
e hidrogel *m*
i idrogel *m*; idrogelo *m*
p hidrogelo *m*
d Hydrogel *n*

1406 hydrogen
f hydrogène *m*
e hidrógeno *m*
i idrogeno *m*
p hidrogênio *m*
d Wasserstoff *m*

1407 hydrogen bomb explosion
f explosion *f* d'une bombe à hydrogène
e explosión *f* de la bomba de hidrógeno
i explosione *f* della bomba all'idrogeno
p explosão *f* da bomba de hidrogênio
d Wasserstoffbombenexplosion *f*

1408 hydrogen sulphide
f sulfure *m* d'hydrogène
e sulfuro *m* de hidrógeno
i sulfuro *m* di idrogeno
p sulfeto *m* de hidrogênio
d Schwefelwasserstoff *m*

1409 hydrographic office
f bureau *m* hydrographique
e instituto *m* hidrográfico
i istituto *m* idrografico
p instituto *m* hidrográfico
d hydrographisches Amt *n*

1410 hydrography
f hydrographie *f*
e hidrografía *f*
i idrografia *f*
p hidrografia *f*
d Hydrographie *f*

1411 hydrological cycle
f cycle *m* de l'eau; circulation *f* de l'eau
e ciclo *m* hidrológico
i ciclo *m* idrologico
p ciclo *m* hidrológico
d Wasserkreislauf *m*

1412 hydrology
f hydrologie *f*
e hidrología *f*
i idrologia *f*
p hidrologia *f*
d Hydrologie *f*

1413 hydrometer
f hydromètre *m*
e hidrómetro *m*
i idrometro *m*
p hidrômetro *m*
d Hydrometer *m*

1414 hydrometer analysis
f analyse *f* aréométrique
e análisis *f* granulométrica con el densímetro
i analisi *f* granulometrica per sedimentazione
p análise *f* hidrométrica
d Schlämmanalyse *f*

1415 hydrometry
f hydrométrie *f*
e hidrometría *f*
i idrometria *f*
p hidrometria *f*
d Hydrometrie *f*

* **hydroponic culture → 1416**

1416 hydroponics; hydroponic culture
f culture *f* hydroponique
e cultivo *m* hidropónico
i idroponica *f*; idrocoltura *f*
p hidropônica *f*; hidrocultura *f*
d Hydroponik *f*; Hydrokultur *f*

1417 hydrosol
f hydrosol *m*
e hidrosol *m*
i idrosol *m*
p hidrossol *m*
d Hydrosol *n*

1418 hydrosphere
f hydrosphère *f*
e hidrosfera *f*
i idrosfera *f*
p hidrosfera *f*
d Hydrosphäre *f*

1419 hydrostatic pressure
f pression *f* hydrostatique
e presión *f* hidrostática
i pressione *f* idrostatica
p pressão *f* hidrostática
d hydrostatischer Druck *m*

1420 hydrostatics
f hydrostatique *f*
e hidrostática *f*
i idrostatica *f*
p hidrostática *f*
d Hydrostatik *f*

1421 hydroxyl
f hydroxyle *m*
e hidróxilo *m*
i ossidrile *m*
p hidróxilo *m*
d Hydroxyl *n*

1422 hyetograph
f hyétographe *m*
e hietografo *m*; mapa *m* pluviométrico
i ietografo *m*
p hietógrafo *m*; mapa *m* pluviométrico
d Hyetograph *m*

1423 hyetometer
f hyétomètre *m*
e hietómetro *m*
i ietometro *m*
p hietômetro *m*
d Hyetometer *n*

1424 hygrograph
f hygrographe *m*
e higrógrafo *m*

i igrografo *m*
p higrógrafo *m*
d Hygrograph *m*

1425 hygrometer
f hygromètre *m*
e higrómetro *m*
i igrometro *m*
p higrômetro *m*
d Hygrometer *n*

1426 hygrometry
f hygrométrie *f*
e higrometría *f*
i igrometria *f*
p higrometria *f*
d Hygrometrie *f*

1427 hygrophyte
f hygrophyte
e higrofita *f*
i igrofita *f*
p higrofita *f*
d Hygrophyt *n*

1428 hygroscope
f hygroscope *m*
e higroscopio *m*
i igroscopio *m*
p higroscópio *m*
d Hygroskop *n*

1429 hygroscopic *adj*
f hygroscopique
e higroscópico
i igroscopico
p higroscópico
d hygroskopisch; wasserziehend

1430 hygroscopic capacity
f capacité *f* hygroscopique
e capacidad *f* higroscópica
i capacità *f* igroscopica
p capacidade *f* higroscópica
d hygroskopische Kapazität *f*

1431 hygroscopicity
f hygroscopicité *f*
e higroscopicidad *f*
i igroscopicità *f*
p higroscopicidade *f*
d Hygroskopizität *f*

1432 hygroscopic water content
f humidité *f* hygroscopique
e humedad *f* higroscópica
i contenuto *m* d'acqua igroscopica
p teor *m* de umidade higroscópica
d Gehalt *m* an hygroskopisch gebundenem Wasser

1433 hyperthermal bath
f bain *m* hyperthermal
e baño *m* hipertermal
i bagno *m* ipertermale
p banho *m* hipertermal
d Überwärmungsbad *n*

1434 hypsometer
f hypsomètre *m*
e hipsómetro *m*
i ipsometro *m*
p hipsômetro *m*
d Hypsometer *n*

1435 hypsometry
f hypsométrie *f*
e hipsometría *f*
i ipsometria *f*
p hipsometria *f*
d Hypsometrie *f*

1436 hysteresis
f hystérésis *f*
e histéresis *f*
i isteresi *f*
p histerese *f*
d Hysterese *f*

I

1437 ice
f glace *f*
e hielo *m*
i ghiaccio *m*
p gelo *m*
d Eis *n*

1438 ice accretion; ice formation; glaciation
f accumulation *f* de glace; dépôt *m* de glace; glaciation *f*
e acumulación *f* de hielo; glaciación *f*
i formazione *f* di ghiaccio; ghiacciamento *m*
p acumulação *f* de gelo; acreção *f* de gelo; glaciação *f*
d Eisansatz *m*; Eisbildung *f*

1439 ice age
f âge *m* glaciaire
e edad *f* glacial
i età *f* glaciale
p idade *f* glacial
d Eiszeit *f*

1440 ice anchor
f ancre *f* à glace
e ancla *f* de hielo
i ancora *f* di ghiaccio
p âncora *f* de gelo
d Eisanker *m*

1441 ice belt
f ceinture *f* de glace
e cinturón *m* de hielo
i cinta *f* di ghiaccio
p cinturão *m* de gelo
d Eisgürtel *m*

1442 iceberg
f iceberg *m*
e iceberg *m*
i iceberg *m*
p icebergue *m*
d Eisberg *m*

1443 iceblink
f reflet *m* éblouissant de la glace; clarté *f* des glaces
e resplandor *m* del hielo; claridad *f* de los hielos
i macchia *f* luminosa nel cielo sopra il ghiaccio
p resplendor *m* de gelo
d Eisblink *n*

1444 ice-bound *adj*
f bloqué par les glaces
e bloqueado por los hielos
i bloccato dai ghiacci
p bloqueado pelos gelos
d vom Eis(e) eingeschlossen

1445 ice-bound harbour
f port *m* bloqué par les glaces
e puerto *m* bloqueado por los hielos
i porto *m* bloccato dai ghiacci
p porto *m* bloqueado pelo gelo
d vom Eis(e) eingeschlossener Hafen *m*

1446 ice bridge
f pont *m* de glace
e puente *f* de hielo
i ponte *f* di ghiaccio galleggiante
p ponte *f* de gelo
d Eisbrücke *f*

* **ice canopy → 2095**

1447 ice cap
f calotte *f* glaciaire
e calota *f* glacial
i calotta *f* glaciale
p calota *f* glacial
d Polkalotte *f*

1448 ice cascade
f cascade *f* de glace
e cascada *f* de hielo
i cascata *f* di ghiaccio
p cascata *f* de gelo
d Eiskaskade *f*

1449 ice chart
f carte *f* des glaces
e carta *f* de los hielos
i carta *f* dei ghiacci
p carta *f* dos gelos
d Eiskarte *f*

1450 ice clause
f clause *f* de glaces
e cláusula *f* de los hielos
i clausola *f* dei ghiacci
p cláusula *f* dos gelos
d Eisklausel *f*

1451 ice conditions
f conditions *fpl* de glace
e condiciones *fpl* de hielo

i condizioni *fpl* di ghiaccio
p condições *fpl* do gelo
d Eisverhältnisse *npl*

1452 ice cooling
f réfrigération *f* par glace
e enfriamiento *m* con hielo
i refrigerazione *f* con ghiaccio idrico
p arrefecimento *m* por gelo
d Eiskühlung *f*

* **ice cover → 1743**

1453 ice-covered island
f île *f* de glace
e isla *f* de hielo
i isola *f* di ghiaccio
p ilha *f* de gelo
d Eisinsel *f*

1454 ice crust; ice rind
f croûte *f* de glace
e crosta *f* de hielo
i crosta *f* di ghiaccio
p crosta *f* de gelo
d Eiskruste *f*; Eishaut *f*

1455 ice crystal
f cristal *m* de glace
e cristal *m* de hielo
i cristallo *m* di ghiaccio
p cristal *m* de gelo
d Eiskristall *n*

1456 ice crystal clouds
f nuages *mpl* contenants des aiguilles de glace
e nubes *fpl* de cristales de hielo
i nubi *fpl* di cristalli di ghiaccio
p nuvens *fpl* compostas de cristais de gelo
d Eiskristallwolken *fpl*

1457 ice day
f jour *m* glacial
e día *m* glacial
i giorno *m* di gelo
p dia *m* glacial
d Eistag *m*

* **ice field → 1458**

1458 ice floe; ice field
f banquise *f*
e banco *m* de hielo; campo *m* de hielo
i banchisa *f*; vedretta *f*
p banco *m* de gelo; campo *m* de gelo
d Eisfeld *n*

1459 ice fog
f brouillard *m* glacé
e niebla *f* de hielo
i nebbia *f* di ghiaccio
p neblina *f* de gelo
d Eisnebel *m*; Frostnebel *m*

* **ice formation → 1438**

1460 ice guard
f antigel *m*
e dispositivo *m* deshelador
i protezione *f* antighiaccio
p proteção *f* antigelo
d Vereisungsschutz *m*

1461 Icelandic low
f minimum *m* islandais; dépression *f* d'Islande
e baja presión *f* de Islandia
i bassa pressione *f* islandese
p sistema *m* anticiclônico situado junto à Islândia
d Island-Tief *n*

1462 Iceland spar
f spath *m* d'Islande
e espato *m* de Islandia
i spato *m* d'Islanda
p espato *m* de Islândia
d isländischer Doppelspat *m*

* **ice layer → 1476**

1463 ice lens
f lentille *f* de glace
e lentejón *m* de hielo
i lente *f* di ghiaccio
p lentícula *f* de gelo
d Eislinse *f*

1464 ice movement
f mouvement *m* de glace
e movimiento *m* de hielo
i movimento *m* di ghiaccio
p movimento *m* de gelo
d Eisbewegung *f*

1465 ice needles
f aiguilles *fpl* de glace
e agujas *fpl* de hielo
i aghi *mpl* di ghiaccio
p agulhas *fpl* de gelo
d freischwebende Eiskristalle *npl*

1466 ice nucleus
f noyau *m* glaçogène
e núcleo *m* de hielo
i nucleo *m* di ghiaccio

p núcleo *m* de gelo
d Eiskern *n*

1467 ice patrol
f patrouille *f* des glaces
e patrulla *f* de hielos
i pattuglia *f* di ghiacci
p patrulha *f* de gelos
d Eiserkundungspatrouille *f*

1468 ice patrol service
f service *m* de patrouilles de glaces
e servicio *m* de patrullas de hielo
i servizio *m* di pattuglie di ghiacci
p serviço *m* de patrulhas de gelo
d Eiswachdienst *m*

1469 ice pellet
f granule *m* de glace; grésil *m*
e gránulo *m* de hielo
i gragnola *f*
p grânulo *m* de gelo
d Eiskörnchen *n*

1470 ice pilot
f pilote *m* des glaces
e práctico *m* de hielos
i pilota *m* di ghiacci
p prático *m* de gelos
d Eislotse *m*

*** ice point → 1125**

1471 ice pressure
f poussée *f* de la glace
e empuje *m* del hielo
i spinta *f* del ghiaccio
p pressão *f* do gelo
d Eisdruck *m*

1472 ice rafting
f transport *m* de glace
e transporte *m* de hielo
i trasporto *m* di ghiaccio
p transporte *m* de gelo
d Eistransport *m*

1473 ice rain
f pluie *f* de glace
e lluvia *f* de hielo; pedrisco *m*
i pioggia *f* ghiacciata
p chuva *f* de granizo
d Eisregen *m*

1474 ice region
f région *f* glaciale
e región *f* glacial
i regione *f* glaciale
p região *f* glacial
d Eisregion *f*

1475 ice reservoir
f réservoir *m* de glace
e reservatorio *m* de hielo
i serbatoio *m* di ghiaccio
p reservatório *m* de gelo
d Eiseinzugsgebiet *n*

*** ice rind → 1454**

1476 ice sheet; ice layer
f couche *f* de glace
e sábana *f* de hielo; cubierta *f* de hielo
i coperta *f* di ghiaccio
p geleira *f* continental; gelo *m* em placas
d Eisschild *m*; Tafeleis *n*

*** ice shelf → 351**

1477 ice stream
f fleuve *m* de glace
e corriente *f* de hielo
i corrente *f* di ghiaccio
p corrente *f* de gelo
d Eisflut *f*; Eisstrom *m*

1478 ice tongue
f langue *f* de glace
e lengua *f* de hielo
i lingua *f* di ghiaccio
p língua *f* de gelo
d Eiszunge *f*

1479 ice water
f eau *f* glacée
e agua *f* helada
i acqua *f* gelata
p áqua *f* gelada
d Eiswasser *n*

1480 icicle
f glaçon *m* stalactite
e cerrión *m*
i ghiacciolo *m*
p sincelos *mpl*; caramelos *mpl* de gelo
d Eiszapfen *m*

1481 icing
f glaçage *m*; givrage *m*
e formación *f* de hielo
i gelata *f*
p formação *f* de gelo
d Vereisung *f*

1482 icing index
f index *m* de givrage
e índice *m* de engelamiento
i indice *m* di formazione del ghiaccio
p índice *m* de formação de gelo
d Vereisungsindex *m*

1483 illuvial horizon
f horizon *m* illuvial
e horizonte *m* iluvial
i orizzonte *m* illuviale
p horizonte *m* iluvial
d Einwaschungshorizont *m*

1484 illuviation
f illuviation *f*
e iluviación *f*
i illuviazione *f*
p iluviação *f*
d Einwaschung *f*

1485 impact
f impact *m*
e impacto *m*
i impatto *m*
p impacto *m*
d Aufprall *m*

1486 impolder *v*; **dam** *v*
f endiguer
e poner diques alrededor
i arginare
p represar; construir um pôlder
d dämmen; einpoldern

1487 in a fog
f en cas de brume
e con niebla
i con nebbia
p com névoa
d bei Nebel

1488 in any sort of weather
f par n'importe quel temps
e con toda clase de tiempo
i per ogni tempo que fa
p com toda sorte de tempo
d bei jedem Wetter

1489 incandescence
f incandescence *f*
e incandescencia *f*
i incandescenza *f*
p incandescência *f*
d Weißglut *f*

1490 inch
f pouce *m*
e pulgada *f*
i pollice *m*
p polegada *f*
d Zoll *m*

1491 inciding ray
f rayon *m* incident
e rayo *m* de caída
i raggio *m* d'incidenza
p raio *m* incidente
d einfallender Strahl *m*

1492 inclination of the wind
f inclinaison *f* du vent
e inclinación *f* del viento
i inclinazione *f* del vento
p inclinação *f* do vento
d Windinklination *f*

1493 incline; rise
f incline *f*; croissance *f*
e pendiente *f*
i incline *m (dello strato)*
p declive *m*
d Anstieg *m*

* **incus → 693**

1494 index
f curseur *m*
e índice *m*
i indice *m*
p índice *m*
d Index *m*

1495 index error; instrument error
f erreur *f* d'index
e error *m* de índice; error *m* de instrumento
i errore *m* d'indice; errore *m* di strumento
p erro *m* de índice
d Indexfehler *m*; Instrumentfehler *m*

1496 Indian Ocean
f Océan *m* indien
e Océano *m* índico
i Oceano *m* indico
p Oceano *m* índico
d indischer Ozean *m*

* **Indian summer → 2293**

1497 induction of flowering
f induction *f* florale
e inducción *f* floral
i induzione *f* a fiore
p indução *f* floral
d Blüteninduktion *f*

1498 industrial waste
f déchet *m* industriel
e residuo *m* industrial
i rifiuto *m* industriale
p resíduo *m* industrial
d Industriemüll *m*

1499 inert gas; noble gas
f gaz *m* inerte; gaz *m* rare
e gas *m* inerte; gas *m* raro
i gas *m* nobile; gas *m* raro
p gás *m* inerte; gás *m* raro
d Inertgas *n*; Edelgas *n*

1500 infertile soil; waste land
f sol *m* stérile
e tierra *f* estéril
i terreno *m* sterile
p solo *m* estéril
d unfruchtbarer Boden *m*

1501 infiltration
f infiltration *f*
e infiltración *f*
i percolazione *f*
p infiltração *f*
d Versickerung *f*

1502 influence of temperature
f influence *f* de la température
e influencia *f* de temperatura
i influsso *m* della temperatura
p influência *f* da temperatura
d Temperatureinfluß *m*

1503 influence on winds
f influence *f* sur les vents
e influencia *f* sobre los vientos
i influenza *f* dei venti
p influência *f* sui venti
d Einfluß *m* auf Winde

1504 infrared *adj*
f infrarouge
e infrarrojo
i infrarosso
p infravermelho
d infrarot

1505 infrared radiation
f rayonnement *m* infrarouge
e radiación *f* infrarroja
i radiazione *f* infrarossa
p radiação *f* infravermelha
d infrarote Strahlung *f*

* **inland ice → 629**

* **inlet → 357**

* **insolation → 2481**

1506 instability
f instabilité *f*
e inestabilidad *f*
i instabilità *f*
p instabilidade *f*
d Instabilität *f*

1507 installations for frost control
f installations *fpl* antigivre
e antiheladas *fpl*
i impianti *mpl* antibrina
p instalações *fpl* anti-geadas
d Frostschutzanlagen *fpl*

* **instrument error → 1495**

1508 instrument meteorological conditions
f conditions *fpl* météorologiques pour vol sans visibilité
e condiciones *fpl* meteorológicas que requeren vuelo por instrumentos
i condizioni *fpl* meteorologiche per volo cieco
p condições *fpl* meteorológicas que requerem vôo por instrumentos
d instrumentenflugbedingende Wetterverhältnisse *npl*

1509 insular climate
f climat *m* insulaire
e clima *m* insular
i clima *m* insulare
p clima *m* insular
d Inselklima *n*

1510 intemperateness
f intempérie *f*
e intemperie *f*
i intemperie *f*
p intempérie *f*
d Wind *m* und Wetter *n*

1511 intensity of gravity
f intensité *f* de la pesanteur
e intensidad *f* de gravidad
i intensità *f* della gravità
p intensidade *f* da gravidade
d Schwereintensität *f*

1512 interception
f captage *m*; interception *f*
e intercepción *f*
i intercettazione *f*
p intercepção *f*
d Abfangen *n*

1513 interglacial *adj*
f interglaciaire
e interglacial
i interglaciale
p interglacial
d interglazial

1514 interglacial epoch
f période *m* interglaciaire; interglaciaire *m*
e época *f* interglacial
i epoca *f* interglaciale
p época *f* interglacial
d Interglazialzeit *f*

1515 interglacial phase
f phase *f* interglaciaire
e fase *f* interglacial
i fase *f* interglaciale
p fase *f* interglacial
d Interglazialphase *f*

1516 intermittent *adj*
f intermittent
e intermitente
i intermittente
p intermitente
d intermittierend

1517 internal constitution
f structure *f* de l'intérieur
e estructura *f* interna
i struttura *f* interna
p estrutura *f* interna
d innerer Aufbau *m*

1518 international ellipsoid
f ellipsoïde *m* international
e elipsoide *m* internacional
i ellissoide *m* internazionale
p elipsóide *m* internacional
d internationales Ellipsoid *n*

1519 international standard atmosphere; ISA
f atmosphère *f* type internationale
e atmósfera *f* tipo internacional
i atmosfera *f* standard internazionale
p atmosfera *f* padrão internacional
d internationale Normalatmosphäre *f*

1520 intertropical front; calm belt; doldrums
f front *m* intertropical; ceinture *f* de calme
e línea *f* fronteriza intertropical; zona *f* de calmas
i zona *f* delle calme; cintura *f* delle calme
p zona *f* de convergência intertropical; cinturão *f* de calmarias
d Kalmengürtel *m*

1521 in the morning; a.m.
f avant midi
e de mañana
i antimeridiano
p de manhã
d vormittags

1522 into the wind; up wind
f face au vent
e contra el viento
i controvento
p contra o vento
d gegen den Wind

1523 inundation; flooding; overflow
f inondation *f*
e inundación *f*; derrame *m*
i inondazione *f*; stramazzo *m*
p inundação *f*; cheia *f*
d Überschwemmung *f*

1524 inversion
f inversion *f*
e inversión *f*
i inversione *f*
p inversão *f*
d Temperaturumkehr *f*

*** in-wintering → 3055**

1525 ion counter; ionization counter
f compteur *m* d'ions
e contador *m* iónico
i contatore *m* a ionizzazione
p contador *m* de íons
d Ionenzähler *m*

*** ionization counter → 1525**

1526 ionogram; ionospheric characteristic
f ionogramme *m*; caractéristique *f* ionosphérique
e ionograma *m*
i ionogramma *m*
p ionograma *m*
d Ionogramm *n*

1527 ionography
f ionographie *f*
e ionografía *f*
i ionografia *f*
p ionografia *f*
d Ionographie *f*

1528 ionosonde
f sondeur *m* ionosphérique
e ionosonda *f*
i ionosonda *f*
p ionosonda *f*
d Ionosonde *f*

1529 ionosphere
f ionosphère *f*
e ionosfera *f*
i ionosfera *f*
p ionosfera *f*
d Ionosphäre *f*

1530 ionospheric *adj*
f ionosphérique
e ionosférico
i ionosferico
p ionosférico
d ionosphärisch

* **ionospheric characteristic → 1526**

1531 ionospheric disturbance
f perturbation *f* ionosphérique
e disturbio *m* ionosférico
i disturbo *m* ionosferico
p distúrbio *m* ionosférico
d Ionosphärenstörung *f*

1532 ionospheric forecast; ionospheric prediction
f prévision *f* ionosphérique
e previsión *f* ionosférica
i previsione *f* ionosferica
p previsão *f* ionosférica
d ionosphärische Vorhersage *f*

1533 ionospheric layer
f couche *f* ionosphérique
e capa *f* ionosférica
i strato *m* ionosferico
p estrato *m* ionosférico
d Ionosphärenschicht *f*

1534 ionospheric observatory
f observatoire *f* ionosphérique
e observatorio *m* ionosférico
i osservatorio *m* ionosferico
p observatório *m* ionosférico
d Ionosphärenobservatorium *n*

* **ionospheric prediction → 1532**

1535 ionospheric region
f région *f* ionosphérique
e región *f* ionosférica
i regione *f* ionosferica
p região *f* ionosférica
d Ionosphärenregion *f*

1536 ionospheric sounding
f sondage *m* ionosphérique
e sondeo *m* ionosférico
i sondaggio *m* ionosferico
p sondagem *f* ionosférica
d Ionosphärenecholotung *f*

1537 ionospheric storm
f orage *m* ionosphérique
e tempestad *f* ionosférica
i burrasca *f* ionosferica
p tempestade *f* ionosférica
d Ionosphärensturm *m*

1538 iridescent cloud
f nuage *m* nacré
e nube *f* irisada
i nube *f* iridescente
p nuvem *f* iridescente
d irisierende Wolke *f*

1539 irregular refraction of rays
f réfraction *f* irrégulière des rayons
e refracción *f* irregular de rayos
i rifrazione *f* irregolare di raggi
p refração *f* irregular de raios
d unregelmäßige Strahlenbrechung *f*

1540 irrigation; watering
f irrigation *f*
e riego *m*; irrigación *f*
i irrigazione *f*
p irrigação *f*
d Bewässerung *f*

1541 irrigation water
f eau *f* d'arrosage
e agua *f* de riego
i acqua *f* irrigua
p água *f* de rega
d Gießwasser *n*

* **ISA → 1519**

1542 isallobar
f isallobare *f*
e isalobara *f*
i isallobara *f*
p isalóbara *f*
d Isallobare *f*

1543 isallobaric field
f aire *f* isallobarique
e campo *m* isalobárico
i campo *m* isallobarico
p área *f* isalobárica
d Fall- und Steiggebiet *n* des Luftdruckes

1544 isallothermic lines
f lignes *fpl* isallothermiques
e isalotérmicos *mpl*
i isallotermiche *mpl*
p isalotérmicos *mpl*
d Isallothermen *fpl*

1545 isanomal; isanomalous line
f isanomale *f*
e línea *f* isoanómala
i isanomala *f*
p linha *f* isanômala
d Isanomale *f*

* **isanomalous line → 1545**

1546 isentropic *adj*
f isentropique
e isentrópico
i isentropico
p isentrópico
d isentropisch

1547 isentropic line
f ligne *f* isentropique
e línea *f* isentrópica
i linea *f* isentropica
p linha *f* isotrópica
d Isentrope *f*

1548 isentropic transformation
f transformation *f* isentropique
e transformación *f* isentrópica
i trasformazione *f* isentropica
p transformação *f* isentrópica
d isentropische Umwandlung *f*

1549 isobar; isobaric line
f isobare *f*; ligne *f* isobare
e isobara *f*; línea *f* isobárica
i isobara *f*; linea *f* isobarica
p isóbara *f*; linha *f* isobárica
d Isobare *f*; isobarische Linie *f*

1550 isobaric chart
f carte *f* isobarique
e carta *f* isobárica
i carta *f* isobarica
p carta *f* isobárica
d Isobarenkarte *f*

* **isobaric line → 1549**

1551 isobaric plane
f surface *f* isobarique
e superficie *f* isobárica
i superficie *f* isobarica
p superfície *f* isobárica
d isobarische Fläche *f*

1552 isobath; isobath line
f isobathe *f*; ligne *f* isobathe
e isobata *f*; línea *f* isobática
i isobata *f*; linea *f* isobatica
p isóbata *f*; linha *f* isobática
d Isobathe *f*; Tiefenlinie *f*

1553 isobath curve
f courbe *f* isobathe
e curva *f* isobática
i curva *f* isobatica
p curva *f* isobática
d Tiefenkurve *f*

* **isobath line → 1552**

1554 isobront
(line on a weather map marking geographical points where thunderstorms occur simultaneously)
f isobronte *f*
e isobronto *m*
i isobronto *m*
p isobronta *f*
d Isobronte *f*

1555 isocheim
(line connecting all points on a weather map having the same mean winter temperature)
f isochimène *f*
e isoquimena *f*
i isochimeno *m*
p isoquímena *f*
d Isochimene *f*

1556 isochrone
(line on a chart or map having the same particular property simultaneously)
f isochrone *f*
e isocrona *f*
i isocrona *f*
p isócrona *f*
d Isochrone *f*

1557 isochronous rolling
f roulis *m* isochrone
e balance *m* isócrono
i rollio *m* isocrono
p balanço *m* isócrono
d gleichzeitiges Schlingern *n*

1558 isoclinal; isoclinic line
f isocline *f*; ligne *f* isoclinale
e isoclina *f*; línea *f* isoclínica
i isoclina *f*; linea *f* isoclina
p isóclino *m*; linha *f* isóclina
d Isokline *f*

1559 isoclinal chart
f carte *f* isocline
e carta *f* isoclina
i carta *f* isoclina
p carta *f* isóclina
d Isoklinenkarte *f*

* **isoclinic line → 1558**

* **isodynamic → 1560**

1560 isodynamic line; isodynamic
f ligne *f* isodynamique; isodyname *f*
e línea *f* isodinámica; isodinámica *f*
i linea *f* isodinamica; isodinamica *f*
p linha *f* isodinâmica; isodinâmica *f*
d isodynamisch Linie *f*; Isodyname *f*

* **isogonal line → 1562**

1561 isogonic chart
f carte *f* isogone
e carta *f* isogónica
i carta *f* isogonica
p carta *f* isogônica
d isogonische Karte *f*

1562 isogonic line; isogonal line
f ligne *f* isogone; isogone *f*
e línea *f* isogona; isogona *f*
i linea *f* isogona; isogona *f*
p linha *f* isogônica; isógona *f*
d Isogone *f*

1563 isogram
f isogramme
e isograma *m*
i isograma *m*
p isograma *m*
d Isogramm *n*

1564 isohaline
(line on a chart or map that connects all points of equal salinity in the ocean)
f isohaline *f*
e isohalina *f*
i isoalina *f*
p isoalina *f*
d Isohaline *f*; Linie *f* gleichen Salzgehaltes

1565 isohel
f isohèle *f*
e isohelia *f*
i isolinea *f* di soleggiamento
p isoélia *f*; linha *f* isoélica
d Isohelie *f*

1566 isohyet
f isohyète *f*
e isohieta *f*
i isoieta *f*
p isoieta *f*
d Isohyete *f*

* **isohypse → 634**

1567 isomer
f isomère *m*
e isómero *m*
i isomero *m*
p isômero *m*
d Isomer *n*

1568 isometric *adj*
f isométrique
e isométrico
i isometrico
p isométrico
d isometrisch

1569 isopleth
f isoplèthe *f*
e isopleta *f*
i isoplete *f*
p isopleta *f*
d Isoplete *f*

1570 isopycnic *adj*
f isopycnique
e isopícnico
i isopicnico
p isopícnico
d isopyknisch

1571 isostasy
f isostasie *f*
e isostasía *f*
i isostasia *f*
p isostasia *f*
d Isostasie *f*

1572 isostatic compensation
f compensation *f* isostatique
e compensación *f* isostática
i compensazione *f* isostatica
p compensação *f* isostática
d isostatische Kompensation *f*

1573 isothere
(imaginary line on a map or chart connecting points having the same mean summer temperature)
f isothère *f*
e isótera *f*
i isotera *f*
p isótera *f*
d Isothere *f*

1574 isotherm; isothermal line
f isotherme *f*; ligne *f* isotherme
e isoterma *f*; línea *f* isoterma
i isoterma *f*; linea *f* isoterma
p isoterma *f*; linha *f* isotérmica
d Isotherme *f*; Linie *f* gleicher Temperatur

1575 isothermal atmosphere
f atmosphère *f* isothermale
e atmósfera *f* isotérmica
i atmosfera *f* isotermica
p atmosfera *f* isotérmica
d isothermische Atmosphäre *f*

1576 isothermal equilibrium
f équilibre *m* isothermique
e equilibrio *m* isotérmico
i equilibrio *m* isotermico
p equilíbrio *m* isotérmico
d isothermes Gleichgewicht *n*

* **isothermal glacier → 2686**

* **isothermal line → 1574**

J

1577 January
f janvier *m*
e enero *m*
i gennaio *m*
p janeiro *m*
d Januar *m*

1578 Japan Current; Kuroshivo
f Courant *m* Kuroshivo
e Corriente *f* de Kuroshivo
i Corrente *f* di Kurosivo
p Corrente *f* de Kuroshivo
d Japanstrom *m*

1579 jauk
(dry, hot foehn wind in the Alps)
f jauk *m*
e jauk *m*
i jauk *m*
p jauk *m*
d Jauk *m*

1580 jet stream
f courant-jet *m*
e corriente *f* de chorro
i corrente *f* a getto
p manga *f* de ar
d Strahlstrom *m*

1581 joule
f joule *m*
e joule *m*
i joule *m*
p joule *m*
d Joule *n*; Wattsekunde *f*

1582 July
f juillet *m*
e julio *m*
i luglio *m*
p julho *m*
d Juli *m*

1583 June
f juin *m*
e junio *m*
i giugno *m*
p junho *m*
d Juni *m*

1584 June drop
f chute *f* de juin
e caída *f* de junio
i cascola *f* di giugno
p queda *f* de junho
d Junifall *m*

K

1585 Kachchan wind
(hot, dry west wind typically of Sri Lanka)
f vent *m* Kachchan
e viento *m* Kachchan
i vento *m* Catchcian
p vento *m* Kachchan
d Kachchanwind *m*

1586 katabatic *adj*
f catabatique
e catabático
i catabatico
p catabático
d katabatisch

1587 katabatic wind; gravity wind; eddying wind; fall-wind
f vent *m* catabatique
e viento *m* catabático
i vento *m* catabatico
p vento *m* catabático
d Fallwind *m*

1588 katathermometer
f catathermomètre *m*
e catatermómetro *m*
i catatermometro *m*
p catatermômetro *m*
d Katathermometer *n*

1589 katisallobar
f catisallobare *f*
e catisalobara *f*
i catisallobara *f*
p catisalóbara *f*
d Katisallobare *f*

1590 khamsin
(dry, hot wind blowing from the Sahara over Egypt and the Red Sea)
f khamsin
e kamsin *m*
i khamsin *m*
p cansim *m*
d Kamsin *m*

1591 kilometric waves
f ondes *fpl* kilométriques
e ondas *fpl* quilométricas
i onde *fpl* chilometriche
p ondas *fpl* quilométricas
d Kilometerwellen *fpl*

1592 kinematic *adj*
f cinématique
e cinemático
i cinematico
p cinemático
d kinematisch

1593 kinetic *adj*
f cinétique
e cinético
i cinetico
p cinético
d kinetisch

1594 Kirchhoff's law
f loi *f* de Kirchhoff
e ley *f* de Kirchhoff
i principio *m* di Kirchhoff
p lei *f* de Kirchhoff
d Kirchhoffsches Strahlungsgesetz *n*

1595 kite
f cerf-volant *m*
e barrilete *m*
i cervo *m* volante
p barrilete *m*
d Drachen *m*

1596 kite balloon
f ballon *m* cerf-volant
e globo-cometa *m*
i pallone *m* frenato
p balão-papagaio *m*
d Drachenballon *n*

1597 knot
f nœud *m*
e nudo *m*
i nodo *m*
p nó *m*
d Knoten *m*

* **Kuroshivo → 1578**

1598 kyle
(tropical storm occurring mainly in the Philippines)
f kyle *m*
e kyle *m*
i kyle *m*
p kyle *m*
d Kyle *m*

L

1599 labile *adj*
f instable; labile
e lábil
i instabile
p lábil; instável
d labil

1600 lability
f labilité *f*
e labilidad *f*
i labilità *f*
p labilidade *f*
d Labilität *f*

1601 Labrador Current
f Courant *m* du Labrador
e Corriente *f* del Labrador
i Corrente *f* del Labrador
p Corrente *f* do Labrador
d Labradorstrom *m*

1602 lag
f décalage *m* temporel
e intervalo *m*
i intervallo *m*
p demora *f*
d zeitliche Verzögerung *f*

1603 lag of tide
f retardation de marée
e retardación *f* de marea
i ritardazione *f* di marea
p atrazo *m* de maré
d Gezeitenverspätung *f*

1604 lake breeze
f brise *f* de lac
e brisa *f* de lago
i brezza *f* di lago
p brisa *f* de lago
d Seebrise *f*

1605 lake ice
f glace *f* de lac
e hielo *m* de lago
i ghiaccio *m* di lago
p gelo *m* de lago
d Seeeis *n*

1606 lambert
(unit of luminance)
f lambert *m*
e lambert *m*
i lambert *m*
p lambert *m*
d Lambert *n*

1607 laminar boundary layer
f couche *f* limite laminaire
e capa *f* límite laminar
i strato *m* limite laminare
p camada *f* limite laminar
d laminare Grenzschicht *f*

1608 laminar flow
f courant *m* laminaire
e corriente *f* laminar; flujo *m* laminar
i corrente *f* laminare
p fluxo *m* laminar
d laminare Strömung *f*

1609 land and sea breezes
f brise *f* de terre et brise *f* de mer
e brisas *fpl* de tierra y de mar
i brezze *fpl* di terra e di mare
p brisas *fpl* da terra e do mar
d Land- und Seewinde *mpl*

1610 land breeze; land wind
f brise *f* de terre; vent *m* de terre
e brisa *f* de tierra; viento *m* da tierra
i brezza *f* di terra; vento *m* da terra
p brisa *f* terrestre; vento *m* da terra
d Landwind *m*

1611 land floe
f glace *f* reliée à la terre
e hielo *m* unido a la tierra
i ghiaccio *m* aderente alla terra
p gelo *m* aderente à terra
d Landeis *n*

1612 land inside the dikes
f terre *f* protégée par des digues
e tierra *f* entre diques
i terra *f* golenale
p terreno *m* entre diques
d Land *n* hinter dem Deich

1613 landslide; earthflow; ground avalanche
f glissement *m* de terrain; avalanche *f* de fond
e corrimiento *m* de tierras; alud *m* de tierra
i smottamento *m*; frana *f*
p deslizamento *m* de terrenos; avalancha *f* de terra
d Erdrutsch *m*; Grundlawine *f*

* **land wind → 1610**

1614 lapse rate
f gradient *m* thermique vertical
e gradiente *m* térmico vertical
i gradiente *m* verticale di temperatura
p gradiente *m* térmico vertical
d vertikaler Temperaturgradient *m*

1615 large-scale growing
f grande culture *f*
e gran cultivo *m*
i grande coltura *f*
p grande cultura *f*
d Großanbau *m*

1616 larval gallery
f dommage *m* causé par un insecte ou un autre phytophage
e superficie *f* dañada
i danno *m* da insetti
p solo *m* danificado
d Fraßstelle *f*

1617 last quarter
f dernier quartier *m*
e cuarto *m* menguante
i ultimo quarto *m*
p quarto *m* minguante
d letztes Viertel *n*

* **LAT → 1664**

1618 late *adj*
f tard
e tarde
i tardo
p tarde
d spät

1619 late frost
f gelée *f* tardive; gel *m* de printemps
e helada *f* tardía
i gelata *f* tardiva
p geada *f* tardia
d Spätfrost *m*

1620 late-maturing *adj*
f de maturation tardive
e de maduración tardía
i di maturazione tardiva
p de maduração tardia
d spätreifend

1621 latent heat
f chaleur *f* latente
e calor *m* latente
i calore *m* latente
p calor *m* latente
d latente Wärme *f*

1622 lateral moraine
f moraine *f* latérale
e morrena *f* lateral
i morena *f* laterale
p morena *f* lateral
d Seitenmoräne *f*

1623 latitude
f latitude *f*
e latitud *f*
i latitudine *f*
p latitude *f*
d Breite *f*

1624 latitude north
f latitude *f* nord
e latitud *f* norte
i latitudine *f* norte
p latitude *f* norte
d nördliche Breite *f*

1625 latitude south
f latitude *f* sud
e latitud *f* sud
i latitudine *f* sud
p latitude *f* sul
d südliche Breite *f*

1626 lava outflow; emission of lava
f coulée *f* de lave
e emisión *f* de lava
i emissione *f* di lava
p emissão *f* de lava
d Ausfluß *m* der Lava

1627 lava plateau
f nappe *f* de lave
e meseta *f* de lava
i plateau *m* di lava
p platô *m* de lava
d Lavadecke *f*

1628 law on nature protection
f loi *f* sur la protection de la nature
e ley *f* sobre la protección de la naturaleza
i legge *f* sulla protezione della natura
p lei *f* de proteção da natureza
d Naturschutzgesetz *n*

1629 leached soil
f sol *m* lessivé
e suelo *m* lavado
i terreno *m* dilavato
p solo *m* lixiviado
d ausgewaschener Boden *m*

1630 leaching; eluviation
f lixiviation *f*
e lixiviación *f*

i lisciviazione *f*
p lixiviação *f*
d Auslaugung *f*

1631 lean-to glasshouse
f serre *f* adossée
e invernadero *m* adosado
i serra *f* murale
p estufa *f* encostada
d Pultdeckgewächshaus *n*

1632 lee tide
f marée *f* sous le vent
e marea *f* de sotavento
i marea *f* di sottovento
p maré *f* de sotavento
d Leegezeiten *fpl*

1633 leeward
f sous le vent
e sotavento
i sottovento
p a sotavento
d unter dem Wind

1634 lee wave
f onde *f* sous le vent
e onda *f* de sotavento
i onda *f* di sottovento
p onda *f* de sotavento
d Leewelle *f*

1635 lenticular cloud
f nuage *m* lenticulaire
e nube *f* lenticular
i nube *f* lenticolare
p nuvem *f* lenticular
d linsenförmige Wolke *f*

1636 leste wind
(hot dry, easterly or southeasterly wind from Canary Islands and Madeira)
f vent *m* leste
e viento *m* leste
i vento *m* leste
p vento *m* leste
d Lestewind *m*

1637 levant
f levant *m*
e levante *m*
i levante *m*
p levante *m*
d Levante *f*

1638 levanter
(strong easterly wind from Mediterranean)
f levante *m*
e levante *m*
i levante *m*
p levante *m*
d starker Südostwind *m*

1639 leveche wind
(hot southern wind occurring in Spain from Sahara)
f vent *m* leveche
e leveche *m*
i vento *m* leveche
p leveche *m*
d Levechewind *m*

1640 light
f lumière *f*
e luz *f*
i luce *f*
p luz *f*
d Licht *n*

1641 light air
f très légère brise *f*
e ventolina *f*
i bava *f* di vento
p ar *m* de baixa pressão
d leiser Zug *m*

1642 light cross
f croix *f* lumineuse
e cruz *f* luminosa
i croce *f* luminosa
p cruz *f* luminosa
d Lichtkreuz *n*

1643 light fog
f brouillard *m* lumineux
e niebla *f* luminosa
i nebbia *f* luminosa
p névoa *f* luminosa
d Lichtnebel *m*

1644 lightning; thunderbolt
f éclair *m*
e relámpago *m*
i lampo *m*
p relâmpago *m*
d Blitz *m*

* **lightning arrester → 1645**

1645 lightning protector; lightning arrester
f parafoudre *m*
e pararrayo(s) *m*
i protettore *m* antifulmini
p pára-raios *m*
d Blitzschutz *m*; Blitzableiter *m*

1646 lightning rod
f tige *f* de paratonnerre
e barra *f* de pararrayo(s)

i parafulmine *m*
p haste *f* de pára-raios
d Blitzableiterstange *f*

1647 light of night sky
f lumière *f* du ciel nocturne
e claridad *f* del cielo nocturno
i luminosità *f* del cielo di notte
p luz *f* celeste noturna
d Licht *n* des Nachthimmels

1648 light ray
f rayon *m* lumineux
e rayo *m* de luz
i raggio *m* di luce
p raio *m* de luz
d Lichtstrahl *m*

1649 light squall
f grainasse *m*
e chubasquillo *m*
i groppo *m* leggiero
p tempestade *f* ligeira
d leichte Bö *f*

* **light wind → 1176**

1650 limit of fog
f limite *f* de la brume
e límite *m* de la niebla
i limite *m* della foschia
p limite *m* da neblina
d Nebelgrenze *f*

1651 limit of trade winds
f limite *f* des vents alizés
e límite *m* de los vientos alisios
i limite *m* dei venti alisei
p limite *m* dos ventos alísios
d Passatgrenze *f*

1652 limnology
f limnologie *f*
e limnología *f*
i limnologia *f*
p limnologia *f*
d Limnologie *f*

1653 line
f ligne *f*
e línea *f*
i linea *f*
p linha *f*
d Linie *f*

1654 line of horizon
f ligne *f* d'horizon
e línea *f* de horizonte
i linea *f* dell'orizzonte
p linha *f* de horizonte
d Horizontlinie *f*

1655 line of pressure
f ligne *f* des pressions
e línea *f* de presiones
i linea *f* di pressioni
p linha *f* de pressões
d Drucklinie *f*

1656 line squall
f ligne *f* de grain
e línea *f* de turbonada
i linea *f* della tempesta; fronte *m* della tempesta
p linha *f* de borrasca; frente *f* fria de uma depressão
d Linienbö *f*; Böenlinie *f*

1657 liquefaction
f liquéfaction *f*
e licuefacción *f*
i liquefazione *f*
p liquefação *f*
d Verflüssigung *f*

1658 liquid
f liquide *m*
e líquido *m*
i liquido *m*
p líquido *m*
d Flüssigkeit *f*

1659 liquid particle; droplet
f particule *f* liquide; gouttelette *f*
e partícula *f* líquida
i particella *f* liquida; gocciolina *f*
p gotícula *f*
d flüssiges Teilchen *n*; Tröpfchen *n*

1660 lithosol
f lithosol *m*; sol *m* squelettique
e suelo *m* esquelético
i litosuolo *m*
p litossolo *m*
d Gesteinsrohboden *m*

1661 lithosphere
f lithosphère *f*
e litosfera *f*
i litosfera *f*
p litosfera *f*
d Lithosphäre *f*

* **little wind → 1176**

1662 littoral current
f courant *m* près de la côte
e corriente *f* costal

i corrente *f* costale
p corrente *f* costeira
d Küstenstrom *m*

1663 livid sky
f ciel *m* gris
e cielo *m* de plomo
i cielo *m* di piombo
p céu *m* lívido
d bleierner Himmel *m*

1664 local apparent time; LAT
f temps *m* apparent local
e hora *f* local aparente
i tempo *m* locale apparente
p hora *f* local verdadeira
d scheinbare Ortszeit *f*

1665 local magnetic disturbance
f perturbation *f* magnétique locale
e perturbación *f* magnética local
i disturbo *m* magnetico locale
p perturbação *f* magnética local
d örtliche magnetische Störung *f*

1666 local time
f heure *f* locale
e hora *f* local
i tempo *m* locale
p hora *f* local
d Ortszeit *f*

1667 local water authority
f administration *f* des eaux
e administración *f* de aguas
i amministrazione *f* delle acque
p autoridade *f* administrativa das águas locais
d Wasserverband *m*

1668 lock
f écluse *f*
e esclusa *f*
i chiusa *f*
p eclusa *f*
d Schleuse *f*

1669 loess
(unstratified calcareous silt formed by fine sand transported by wind)
f loess *m*
e loess *m*
i loess *m*
p loess *m*
d Löß *m*

1670 longitude
f longitude *f*
e longitud *f*
i longitudine *f*
p longitudine *f*
d Länge *f*

1671 longitudinal wave
f onde *f* longitudinale
e onda *f* longitudinal
i onda *f* longitudinale
p onda *f* longitudinal
d Longitudinalwelle *f*

1672 long-period forecast; long-range forecast
f prévision *f* à longue échéance
e pronóstico *m* a largo plazo
i previsione *f* a lunga scadenza
p prognóstico *m* de longo alcance
d langfristige Prognose *f*; Langfristprognose *f*

* **long-range forecast → 1672**

1673 long wave
f onde *f* longue
e onda *f* larga
i onda *f* lunga
p onda *f* longa
d Langwelle *f*

1674 looming
f déplacement *m* de l'horizon
e espejismo *m* de calina
i ingrandimento *m* apparente
p amplificação *f* aparente
d Sichtbarwerden *n*

1675 low clouds
f nuages *mpl* bas
e nubes *fpl* bajas
i nubi *fpl* basse
p nuvens *fpl* baixas
d tiefe Wolken *fpl*

1676 lower atmosphere
f atmosphère *f* inférieure
e atmósfera *f* inferior
i atmosfera *f* inferiore
p atmosfera *f* inferior
d untere Atmosphäre *f*

1677 lower parhelion
f parhélie *m* inférieur
e parhelio *m* inferior
i parelio *m* inferiore
p parélio *m* inferior
d Untersonne *f*

1678 lower tangential arc
f arc *m* tangent inférieur
e arco *m* tangente inferior

i arco *m* tangente inferiore
p arco *m* tangente inferior
d unterer Berührungsbogen *m*

1679 low frequency
f basse fréquence *f*
e baja frecuencia *f*
i bassa frequenza *f*
p baixa frequência *f*
d niedrige Frequenz *f*

1680 low latitudes
f latitudes *fpl* basses
e latitudes *fpl* bajas
i latitudini *fpl* basse
p latitudes *fpl* baixas
d niedrige Breiten *fpl*

1681 low pressure area
f zone *f* de basse pression
e zona *f* de depresión barométrica; área *f* de bajas presiones
i zona *f* di bassa pressione
p zona *f* de baixa pressão
d Tiefdruckgebiet *n*

1682 low temperature
f basse température *f*
e baja temperatura *f*
i bassa temperatura *f*
p baixa temperatura *f*
d tiefe Temperatur *f*

1683 low-temperature breakdown
f brunissement *m* interne par des basses températures
e descomposición *f* interna devida a bajas temperaturas
i disfacimento *m* da bassa temperatura
p acastanhamento *m* interno devido a baixas temperaturas
d Kältefleischbräune *f*

1684 low-temperature stimulus
f effet *m* stimulateur du froid
e estímulo *m* a baja temperatura
i stimolo *m* da freddo
p estímulo *m* à baixa temperatura
d Kältereiz *m*

1685 low visibility; poor visibility
f mauvaise visibilité *f*
e mala visibilidad *f*
i visibilità *f* cattiva
p má visibilidade *f*
d geringe Sichtigkeit *f*

* **low water → 855**

1686 low-water full and change
f établissement *m* de basse mer
e establecimiento *m* de bajamar
i stabilimento *m* volgare per le basse maree
p estabelecimento *m* de baixa-mar
d mittleres Niedrigwasserintervall *n*

1687 low-water level
f niveau *m* de la marée basse
e nivel *m* de bajamar
i livello *m* della bassa marea
p nível *m* da baixa-mar
d Niedrigwasserebene *f*

1688 low-water quadrature
f intervalle *m* de la basse mer aux quadratures
e intervalo *m* de la bajamar en cuadratura
i quadratura *f* di acqua bassa
p quadratura *f* da baixa-mar
d quadraturale Niedrigwasserintervall *n*

1689 luminescence; glow
f luminescence *f*; faible lueur *f*; lueur *f*
e luminiscencia *f*
i luminescenza *f*; debole luminosità *f*
p luminescência *f*
d Lumineszenz *f*; Leuchten *n*

* **luminosity → 424**

1690 luminous *adj*
f lumineux
e luminoso
i luminoso
p luminoso
d leuchtend

1691 luminous clouds
f nuages *mpl* lumineux
e nubes *fpl* luminosas
i nubi *fpl* luminose
p nuvens *fpl* luminosas
d leuchtende Nachtwolken *fpl*

* **luminous coefficient → 1692**

1692 luminous efficiency; luminous coefficient
f éclairement *m* lumineux relatif
e coeficiente *m* luminoso
i coefficiente *m* luminoso
p coeficiente *m* luminoso
d Lichtausbeute *f*

1693 luminous flux
f flux *m* lumineux
e flujo *m* luminoso
i flusso *m* luminoso
p fluxo *m* luminoso
d Lichtstrom *m*

1694 luminous intensity
f intensité *f* lumineuse
e intensidad *f* luminosa
i intensità *f* luminosa
p intensidade *f* luminosa
d Beleuchtungsstärke *f*

1695 lunar *adj*
f lunaire
e lunar
i lunare
p lunar
d lunar

1696 lunar atmospheric tide
f marée *f* atmosphérique lunaire
e marea *f* atmosférica lunar
i marea *f* atmosferica lunare
p maré *f* atmosférica lunar
d lunare Atmosphärengezeiten *fpl*

1697 lunar aureole; lunar corona
f auréole *f* lunaire; couronne *f* lunaire
e corona *f* lunar
i aureola *f* lunare; corona *f* lunare
p auréola *f* lunar
d Mondkranz *m*

1698 lunar circle
f cercle *m* lunaire
e círculo *m* lunar
i circolo *m* lunare
p círculo *m* lunar
d Mondkreis *m*

*** lunar corona → 1697**

1699 lunar cross
f croix *f* lunaire
e cruz *f* lunar
i croce *f* lunare
p cruz *f* lunar
d Mondkreuz *n*

1700 lunar cycle
f cycle *m* lunaire
e ciclo *m* lunar
i ciclo *m* lunare
p ciclo *m* lunar
d Mondzyklus *m*

1701 lunar day
f jour *m* lunaire
e día *m* lunar
i giorno *m* lunare
p dia *m* lunar
d Mondtag *m*

1702 lunar distance
f distance *f* lunaire
e distancia *f* lunar
i distanza *f* lunare
p distância *f* lunar
d Monddistanz *f*

1703 lunar eclipse
f éclipse *f* de la lune
e eclipse *m* lunar
i eclisse *f* lunare
p eclipse *m* lunar
d Mondfinsternis *f*

1704 lunar halo
f halo *m* lunaire
e halo *m* lunar
i alone *m* lunare
p halo *m* lunar
d Mondring *m*

1705 lunar hour
f heure *f* lunaire
e hora *f* lunar
i ora *f* lunare
p hora *f* lunar
d Mondstunde *f*

1706 lunar meridian transit
f passage *m* de la lune au méridien
e pasaje *f* de la luna en el meridiano
i passaggio *m* della luna nel meridiano
p passagem *f* da lua pelo meridiano
d Mondmeridiandurchgang *m*

1707 lunar month
f mois *m* lunaire
e mes *m* lunar
i mese *m* lunare
p mês *m* lunar
d Mondmonat *m*

1708 lunar orbit
f orbite *f* de la lune
e órbita *f* lunar
i orbita *f* lunare
p órbita *f* lunar
d Mondbahn *f*

1709 lunar year
f année *f* lunaire
e año *m* lunar
i anno *m* lunare
p ano *m* lunar
d Mondjahr *n*

1710 lunation
f lunaison *f*
e lunación *f*
i lunazione *f*

p lunação *f*
d Lunation *f*

1711 lunitidal interval
f intervalle *m* lunaire
e intervalo *m* luna-marea
i intervallo *m* luna-marea
p intervalo *m* lua-maré
d Hochwasserintervall *n*

1712 lux
(unit of illumination)
f lux *m*
e lux *m*
i lux *m*
p lux *m*
d Lux *n*

1713 luxmeter
f luxmètre *m*
e medidor *m* de radiación
i luxmetro *m*
p lucímetro *m*
d Beleuchtungsmesser *m*

1714 lysimeter
f lysimètre *m*
e lisímetro *m*
i lisimetro *m*
p lisímetro *m*
d Lysimeter *n*

M

1715 mackerel sky
f ciel *m* moutonné; ciel *m* pommelé
e cielo *m* con nubes onduladas
i cielo *m* con nubi ondulante
p céu *m* encarneirado
d Schäfchenwolkenhimmel *m*

1716 macroclimate
f macroclimat *m*
e macroclima *m*
i macroclima *m*
p macroclima *m*
d Großraumklima *n*

1717 magma
f magma *m*
e magma *m*
i magma *m*
p magma *f*
d Magma *n*

1718 magnetic *adj*
f magnétique
e magnético
i magnetico
p magnético
d magnetisch

1719 magnetic axis
f axe *m* magnétique
e eje *m* magnético
i asse *m* magnetico
p eixo *m* magnético
d magnetische Achse *f*

1720 magnetic bearing
f relèvement *m* magnétique
e marcación *f* magnética
i rilevamento *m* magnetico
p marcação *f* magnética
d magnetische Peilung *f*

1721 magnetic chart
f carte *f* magnétique
e carta *f* magnética
i carta *f* magnetica
p carta *f* magnética
d magnetische Karte *f*

1722 magnetic declination
f déclinaison *f* magnétique
e variación *f* magnética
i declinazione *f* magnetica
p declinação *f* magnética
d magnetische Deklination *f*

1723 magnetic disturbance
f perturbation *f* magnétique
e perturbación *f* magnética
i disturbo *m* magnetico
p perturbação *f* magnética
d magnetische Störung *f*

1724 magnetic equator
f équateur *m* magnétique
e ecuador *m* magnético
i equatore *m* magnetico
p equador *m* magnético
d magnetischer Äquator *m*

1725 magnetic field
f champ *m* magnétique
e campo *m* magnético
i campo *m* magnetico
p campo *m* magnético
d Magnetfeld *n*

1726 magnetic inclination
f inclinaison *f* magnétique
e inclinación *f* magnética
i inclinazione *f* magnetica
p inclinação *f* magnética
d magnetische Inklination *f*

1727 magnetic meridian
f méridien *m* magnétique
e meridiano *m* magnético
i meridiano *m* magnetico
p meridiano *m* magnético
d magnetischer Meridian *m*

1728 magnetic midnight
f minuit *m* magnétique
e medianoche *f* magnética
i mezzanotte *f* magnetica
p meia-noite *f* magnética
d magnetische Mitternacht *f*

1729 magnetic moment
f moment *m* magnétique
e momento *m* magnético
i momento *m* magnetico
p momento *m* magnético
d magnetisches Moment *n*

1730 magnetic needle
f aiguille *f* aimantée
e aguja *f* magnética

i ago *m* magnetico
p agulha *f* magnética
d Magnetnadel *f*

1731 magnetic north
f nord *m* magnétique
e norte *m* magnético
i nord *m* magnetico
p norte *m* magnético
d mißweisender Nord *m*

1732 magnetic pole
f pôle *m* magnétique
e polo *m* magnético
i polo *m* magnetico
p pólo *m* magnético
d Magnetpol *m*

* **magnetic pole of the earth → 885**

1733 magnetic storm
f tempête *f* magnétique
e tempestad *f* magnética
i tempesta *f* magnetica
p tempestade *f* magnética
d magnetischer Sturm *m*

1734 magnetism
f magnétisme *m*
e magnetismo *m*
i magnetismo *m*
p magnetismo *m*
d Magnetismus *m*

1735 magnetograph
f magnétographe *m*
e magnetógrafo *m*
i magnetografo *m*
p magnetógrafo *m*
d Magnetograph *m*

1736 magnetometer
f magnétomètre *m*
e magnetómetro *m*
i magnetometro *m*
p magnetômetro *m*
d Magnetometer *n*

1737 make *v* **serene**
f rasséréner
e serenar
i rasserenare
p serenar
d aufklären

1738 maloja
(wind from mountain and from valley in Switzerland)
f maloja *m*
e maloja *m*
i maloia *m*
p maloja *m*
d Malojawind *m*

1739 mammatocumulus
(cloud characterized by breast-shaped protuberances along the lower surface)
f mammato-cumulus *m*
e mamatocumulus *m*
i mammatocumulo *m*
p mamatocúmulo *m*
d Mammatokumulus *m*

1740 mammatus
(cloud whose lower surface is characterized by appearance with pouches)
f mammatus *m*
e mamatus *m*
i mammato *m*
p mamato *m*
d Mammatus *m*

1741 manipulation of flowering
f influence *f* sur la floraison
e influencia *f* a floración
i influenzamento *m* fiorale
p influência *f* sobre a floração
d Blühbeeinflussung *f*

1742 manometer
f manomètre *m*
e manómetro *m*
i manometro *m*
p manômetro *m*
d Manometer *n*

1743 mantle ice; ice cover
f glace *f* de couverture
e manto *m* de hielo
i ghiaccio *m* del mantello; concentrazione *f* del ghiaccio
p manto *m* de gelo
d Eisdecke *f*

1744 March
f mars *m*
e marzo *m*
i marzo *m*
p março *m*
d März *m*

1745 mare
f mer *f* lunaire
e mar *m* lunar
i mare *m* lunare
p maré *f* lunar
d Mondmare *f*

1746 mare's tail
(cirrus cloud appearing the tail of a horse)
f cirrus *m* en queue de cheval
e rabo *m* de gallo
i coda *f* di cavallo
p cauda *f* de cavalo
d Federwolke *f*

1747 marigraph; tide gauge
f marégraphe *m*
e mareógrafo *m*
i mareografo *m*
p mareógrafo *m*
d Mareograph *m*

1748 marine barometer
f baromètre *m* de marine
e barómetro *m* marino
i barometro *m* marino
p barômetro *m* de marinha
d Marineluftdruckmesser *m*

1749 marine ecology
f écologie *f* marine
e ecología *f* marina
i ecologia *f* marina
p ecologia *f* marinha
d marine Ökologie *f*

1750 marine glaciology
f glaciologie *f* marine
e glaciología *f* marina
i glaciologia *f* marina
p glaciologia *f* marinha
d marine Glaziologie *f*

1751 maritime air
f air *m* maritime
e aire *m* marítimo
i aria *f* marittima
p ar *m* marítimo
d Seeluft *f*

1752 maritime climate
f climat *m* maritime
e clima *m* marítimo
i clima *m* marittimo
p clima *m* marítimo
d Küstenklima *n*; Seeklima *n*

1753 maritime weather office
f service *m* météorologique maritime
e servicio *m* meteorológico marítimo
i servizio *m* meteorologico marittimo
p serviço *m* meteorológico marítimo
d Seewetteramt *n*

1754 marsh; swamp
f marais *m*
e estero *m*; terreno *m* pantanoso
i palude *f*; acquitrino *m*
p priel *m*; pântano *m*
d Moor *n*

1755 marshy ground
f terrain *m* marécageux
e terreno *m* paludoso
i terreno *m* paludoso
p terreno *m* paludoso
d Sumpfboden *m*

1756 mass movement
f mouvement *m* de masse
e movimiento *m* de masa
i movimento *m* di massa
p movimento *m* de massa
d Massenbewegung *f*

1757 maturing
f vieillissement *m*
e maduración *f*
i invecchiamento *m*
p envelhecimento *m*
d Lagerung *f*

1758 maximum and minimum thermometer
f thermomètre *m* à maxima et minima
e termómetro *m* de máxima y mínima
i termometro *m* a massima e minima
p termômetro *m* de máxima e mínima
d Maximum- und Minimumthermometer *n*

1759 maximum dry density
f densité *f* sèche maximale
e densidad *f* seca máxima
i peso *m* massimo dell'unità di volume della terra essiccata
p máxima baridade *f* seca
d maximales Trockenraumgewicht *n*

1760 maximum temperature
f température *f* maximum
e temperatura *f* máxima
i temperatura *f* massima
p temperatura *f* máxima
d Maximaltemperatur *f*

1761 maximum thermometer
f thermomètre *m* à maxima
e termómetro *m* de máxima
i termometro *m* a massima
p termômetro *m* de máxima
d Maximalthermometer *n*

1762 May
f mai *m*
e mayo *m*
i maggio *m*
p maio *m*
d Mai *m*

1763 mean; average
f moyenne *f*
e media *f*; promedio *m*
i media *f*
p média *f*
d Mittel *n*

1764 mean low water
f marée *f* basse moyenne
e marea *f* baja media
i bassa marea *f* media
p maré *f* baixa média
d mittleres Niedrigwasser *n*

1765 mean sea level
f niveau *m* moyen de la mer
e nivel *m* medio del mar
i livello *m* medio del mare
p nível *m* médio do mar
d mittlerer Seestand *m*

1766 mean solar day
f jour *m* solaire moyen
e día *m* medio solar
i giorno *m* solare medio
p dia *m* médio
d mittlerer Sonnentag *m*

1767 mean sun
f soleil *m* moyen
e sol *m* medio
i sole *m* medio
p sol *m* médio
d mittlere Sonne *f*

1768 mean tide level
f niveau *m* moyen de la marée
e nivel *m* medio de la marea
i livello *m* medio della marea
p nível *m* médio da maré
d Mittelwasser *n*

1769 mean time
f temps *m* moyen
e tiempo *m* medio
i tempo *m* medio
p tempo *m* médio
d mittlere Zeit *f*

1770 mean value; average value
f valeur *f* moyenne
e valor *m* medio
i valor *m* medio
p valor *m* médio
d Durchschnittswert *m*

1771 mean water level
f niveau *m* moyen de l'eau
e nivel *m* medio del agua
i livello *m* medio dell'acqua
p nível *m* médio da água
d mittlerer Wasserstand *m*

1772 measurement of a degree
f mesure *f* du degré
e medición *f* de grados
i misurazione *f* del grado
p medição *f* do grau
d Gradmessung *f*

1773 measurement of temperature
f mesure *f* de la température
e medición *f* de la temperatura
i misura *f* della temperatura
p medição *f* da temperatura
d Temperaturmessung *f*

1774 measure of time
f mesure *f* de temps
e medida *f* de tiempo
i misura *f* di tempo
p medida *f* de tempo
d Zeitmaß *n*

1775 mechanical equivalent of heat
f équivalent *m* mécanique de la chaleur
e equivalente *m* mecánico del calor
i equivalente *m* meccanico della caloria
p equivalente *m* mecânico do calor
d Arbeitswert *m* der Wärme

* **mechanical erosion → 649**

* **mechanics of ice → 868**

1776 medial moraine
f moraine *f* médiane
e morrena *f* media
i morena *f* mediana
p morena *f* média
d Mittelmoräne *f*

1777 medium clouds
f nuages *mpl* moyens
e nubes *fpl* medias
i nubi *fpl* medie
p nuvens *fpl* médias
d mittelhohe Wolken *fpl*

1778 medium frequency
f fréquence *f* moyenne
e frecuencia *f* media
i frequenza *f* media
p frequência *f* média
d mittlere Frequenz *f*

1779 medium wave
f onde *f* moyenne
e onda *f* media
i onda *f* media
p onda *f* média
d Mittelwelle *f*

1780 melt *v*
f fondre
e fundir
i fondere
p fundir
d flüssigen

1781 melted snow and ice
f eau *f* de fusion de la glace
e agua *f* de fusión de hielo
i acqua *f* di fusione del ghiaccio
p água *f* de fusão do gelo
d Schmelzwasser *n*

1782 melting point
f point *m* de fusion
e punto *m* de fusión
i punto *m* di fusione
p ponto *m* de fusão
d Schmelzpunkt *m*

1783 meniscus
f ménisque *m*
e menisco *m*
i menisco *m*
p menisco *m*
d Meniskus *m*

1784 mercury barometer
f baromètre *m* à mercure
e barómetro *m* de mercurio
i barometro *m* a mercurio
p barômetro *m* de mercúrio
d Quecksilberbarometer *n*

1785 mercury gauge
f manomètre *m* à mercure
e manómetro *m* de mercurio
i manometro *m* di mercurio
p manômetro *m* de mercúrio
d Quecksilbersäule *f*

1786 mercury thermometer
f thermomètre *m* à mercure
e termómetro *m* de mercurio
i termometro *m* a mercurio
p termômetro *m* de mercúrio
d Quecksilberthermometer *n*

1787 meridian
f méridien *m*
e meridiano *m*
i meridiano *m*
p meridiano *m*
d Meridian *m*; Mittagskreis *m*

1788 meridional *adj*
f méridional
e meridional
i meridionale
p meridional
d meridional

1789 meridional altitude of the sun
f hauteur *f* méridienne du soleil
e altura *f* meridiana del sol
i altezza *f* meridiana del sole
p altura *f* meridiana do sol
d Mittagshöhe *f* der Sonne

1790 meridional distance
f différence *f* de longitude
e diferencia *f* de longitud
i differenza *f* di longitudine
p diferença *f* de longitude
d Längenunterschied *m*

1791 mesoclimate
f mésoclimat *m*
e mesoclima *m*
i mesoclima *m*
p mesoclima *m*
d Ortsklima *n*

1792 mesopause
f mésopause *f*
e mesopausa *f*
i mesopausa *f*
p mesopausa *f*
d Mesopause *f*

1793 mesophyll
f mésophylle *m*
e mesófilo *m*
i mesofillo *m*
p mesófilo *m*
d Blattgewebe *n*

1794 mesosphere
f mésosphère *f*
e mesoesfera *f*
i mesosfera *f*
p mesosfera *f*
d Mesosphäre *f*

1795 metamorphism of ice
f métamorphisme *m* de la glace
e metamorfismo *m* del hielo
i metamorfismo *m* del ghiaccio
p metamorfismo *m* do gelo
d Eismetamorphose *f*

1796 metamorphism of snow
f métamorphisme *m* de la neige
e metamorfismo *m* de la nieve
i metamorfismo *m* della neve
p metamorfismo *m* da neve
d Schneemetamorphose *f*

1797 meteor
f météore *m*
e meteoro *m*
i meteora *f*; stella *f* cadente
p meteoro *m*
d Meteor *m*

1798 meteoric *adj*
f météorique
e meteórico
i meteorico
p meteórico
d meteorisch

1799 meteoric water
f eau *f* météorique
e agua *f* meteórica
i acqua *f* meteorica
p água *f* meteórica
d meteorisches Wasser *n*

1800 meteorism
f météorisme *m*
e meteorismo *m*
i meteorismo *m*
p meteorismo *m*
d Meteorismus *m*

1801 meteorite
f météorite *f*
e meteorito *m*
i meteorite *f*
p meteorito *m*
d Meteorit *m*

* **meteorograph → 349**

1802 meteorological *adj*
f météorologique
e meteorológico
i meteorologico
p meteorológico
d meteorologisch

1803 meteorological aeroplane
f avion *m* pour les services de météorologie
e avión *m* para servicios meteorológicos
i aeroplano *m* per servizi meteorologici
p avião *m* para serviços meteorológicos
d Wetterflugzeug *m*

1804 meteorological chart; weather chart
f carte *f* météorologique
e carta *f* meteorológica
i carta *f* meteorologica
p carta *f* meteorológica
d Wetterkarte *f*

* **meteorological conditions → 2965**

1805 meteorological effect
f influence *f* météorologique
e influencia *f* meteorológica
i variazione *f* meteorologica
p efeito *m* meteorológico
d Wetterwirkung *f*

1806 meteorological element
f élément *m* météorologique
e elemento *m* meteorológico
i elemento *m* meteorologico
p elemento *m* meteorológico
d meteorologisches Element *n*

1807 meteorological information during flight
f avertissement *m* météorologique en vol
e información *f* meteorológica en vuelo
i informazione *f* meteorologica in volo
p informação *f* meteorológica em vôo
d Flugwetterberatung *f*

1808 meteorological instrument
f instrument *m* météorologique
e instrumento *m* meteorológico
i strumento *m* meteorologico
p instrumento *m* meteorológico
d meteorologisches Instrument *n*

1809 meteorological journal
f journal *m* météorologique
e gaceta *f* meteorológica
i giornale *m* meteorologico
p jornal *m* meteorológico
d meteorologisches Tagebuch *n*

1810 meteorological navigation
f navigation *f* météorologique
e navegación *f* meteorológica
i navigazione *f* meteorologica
p navegação *f* meteorológica
d Wetternavigation *f*

* **meteorological observatory → 2978**

1811 meteorological optical range
f portée *f* optique météorologique
e alcance *m* óptico meteorológico
i portata *f* ottica meteorologica
p alcance *m* ótico meteorológico
d meteorologische optische Sichtweite *f*

1812 meteorological protection
f protection *f* météorologique
e protección *f* meteorológica
i protezione *f* meteorologica
p proteção *f* meteorológica
d Wetterprotektion *f*

* **meteorological report → 2975**

1813 meteorological service
f service *m* météorologique
e servicio *m* meteorológico
i servizio *m* meteorologico
p serviço *m* meteorológico
d Wetterdienst *m*

1814 meteorological sign
f signe *m* de météorologie
e signo *m* meteorológico
i segno *m* meteorologico
p sinal *m* meteorológico
d meteorologiches Zeichen *n*

1815 meteorological stability
f stabilité *f* météorologique
e estabilidad *f* meteorológica
i stabilità *f* meteorologica
p estabilidade *f* meteorológica
d meteorologische Stabilität *f*

1816 meteorological survey
f observation *f* météorologique
e observación *f* meteorológica
i osservazione *f* meteorologica
p observação *f* meteorológica
d Wetterbeobachtung *f*

1817 meteorological symbol
f symbole *m* météorologique
e símbolo *m* meteorológico
i simbolo *m* meteorologico
p símbolo *m* meteorológico
d Wetterzeichen *n*

1818 meteorologist
f météorologiste *m*
e meteorólogo *m*
i meteorologo *m*
p meteorólogo *m*
d Meteorologe *m*

1819 meteorology
f météorologie *f*
e meteorología *f*
i meteorologia *f*
p meteorologia *f*
d Meteorologie *f*

1820 meteoropathology
f météoropathologie *f*
e meteoropatología *f*
i meteoropatologia *f*
p meteoropatologia *f*
d Meteoropathologie *f*

1821 meteorotropism
f météorotropisme *m*
e meteorotropismo *m*
i meteorotropismo *m*
p meteorotropismo *m*
d Meteorotropismus *m*

1822 methane
f méthane *m*
e metano *m*
i metano *m*
p metano *m*
d Methan *n*

1823 microbarograph
f microbarographe *m*
e microbarógrafo *m*
i microbarografo *m*
p microbarógrafo *m*
d Mikrobarograph *m*

1824 microclimate
f microclimat *m*
e microclima *m*
i microclima *m*
p microclima *m*
d Mikroklima *n*

1825 microflora
f microflore *f*
e microflora *f*
i microflora *f*
p microflora *f*
d Mikroflora *f*

1826 microseism
f microséisme *m*
e microsismo *m*
i microsismo *m*
p microssismo *m*
d leichtes Erdbeben *n*

1827 midday
f midi *m*
e mediodía *m*
i mezzogiorno *m*
p meio-dia *m*
d Mittag *m*

1828 middle atmosphere
f atmosphère *f* moyenne
e atmósfera *f* media

i atmosfera *f* media
p atmosfera *f* média
d mittlere Atmosphäre *f*

1829 midnight
f minuit *m*
e medianoche *f*
i mezzanotte *f*
p meia-noite *f*
d Mitternacht *f*

1830 migration of dunes
f migration *f* des dunes
e migración *f* de dunas
i migrazione *f* di dune
p migração *f* de dunas
d Wandern *n* der Dünen

1831 mild climate
f climat *m* doux
e clima *m* dulce
i clima *m* dolce
p clima *m* doce
d mildes Klima *n*

1832 mild weather
f temps *m* doux
e tiempo *m* suave
i tempo *m* dolce
p tempo *m* doce
d mildes Wetter *n*

1833 mile
f mille *m*
e milla *f*
i miglio *m*
p milha *f*
d Meile *f*

1834 millibar
(unit of pressure equivalent to 0.001 bar)
f millibar *m*
e milibar *m*
i millibar *m*
p milibar *m*
d Millibar *n*

1835 millibar scale
f graduation *f* en millibars
e escala *f* en milibares
i scala *f* en millibari
p escala *f* em milibares
d Millibarteilung *f*

1836 minauroral belt
f ceinture *f* minaurorale
e cinturón *m* minauroral
i fascia *f* minaurorale
p cinturão *m* minauroral
d Gürtel *m* geringer Polarlichthäufigkeit

1837 mine barometer
f baromètre *m* de mine
e barómetro *m* de mina
i barometro *m* di miniera
p barômetro *m* de mina
d Grubenbarometer *n*

1838 minimum temperature
f température *f* minimum
e temperatura *f* mínima
i temperatura *f* minima
p temperatura *f* mínima
d Minimaltemperatur *f*

1839 minimum thermometer
f thermomètre *m* à minima
e termómetro *m* de mínima
i termometro *m* a minima
p termômetro *m* de mínimas
d Minimalthermometer *n*

1840 minuano
(a cold, dry winter wind blowing from southern Brazil)
f minuano *m*
e minuano *m*
i minuano *m*
p minuano *m*
d Minuano *m*

1841 mirage
f mirage *m*
e espejismo *m*
i miraggio *m*
p miragem *f*
d Luftspiegelung *f*; Fata Morgana *f*

1842 mist
f brume *f*
e neblina *f*
i foschia *f*
p neblina *f*
d feuchter Dunst *m*

1843 mist condition
f condition *f* de brume
e condición *f* de niebla
i condizione *f* di nebbia
p condição *f* de neblina
d nebelförmiger Zustand *m*

1844 mistral; mistral wind
(strong, cold dry northerly wind in the Gulf of Lyon and lower Adriatic)
f mistral *m*
e mistral *m*

i mistrale *m*
p mistral *m*
d Mistral *m*

* **mistral wind → 1844**

* **misty** *adj* **→ 1320**

1845 mixed cropping
f cultures *fpl* associées
e cultivo *m* mixto
i produzione *f* mista
p cultura *f* mista
d Mischkultur *f*

1846 mixed tides
f marées *fpl* mixtes
e mareas *fpl* mixtas
i maree *fpl* miste
p marés *fpl* mistas
d gemischte Gezeiten *fpl*

* **mock sun → 2019**

1847 moderate *adj*
f modéré
e moderado
i moderato
p moderado
d mäßig

1848 moderate breeze
f jolie brise *f*
e brisa *f* moderada
i brezza *f* moderata
p brisa *f* moderada
d mäßige Brise *f*

1849 moderate gale
f grand grais *m*
e viento *m* muy fresco
i burrasca *f* moderata
p vento *m* muito fresco
d steife Brise *f*

1850 module
f module *m*; modulus *m*
e módulo *m*
i modulo *m*
p módulo *m*
d Modul *m*

1851 moist *adj*; **damp** *adj*
f humide
e húmedo
i umido
p úmido
d feucht

1852 moist air; humid air; damp air
f air *m* humide
e aire *m* húmedo
i aria *f* umida
p ar *m* úmido
d feuchte Luft *f*

1853 moist curing
f humidification *f*
e curación *f* húmeda
i umidificazione *f*
p cura *f* por umidade
d Feuchthaltung *m*

1854 moistening
f humidification *f*
e humectación *f*
i inumidimento *m*
p umectação *f*
d Anfeuchten *n*

1855 moist meadow
f prairie *f* humide
e pradera *f* húmeda
i prato *m* umido
p prado *m* úmido
d Feuchtwiese *f*

* **moisture → 1396**

1856 moisture absorption
f absorption *f* d'humidité
e absorción *f* de humedad
i assorbimento *m* di umidità
p absorção *f* de umidade
d Feuchtigkeitsabsorption *f*

1857 moisture content
f teneur *f* en humidité; contenu *m* en vapeur d'eau
e contenido *m* de humedad
i tenore *m* d'acqua
p teor *m* de umidade
d Feuchtigkeitsgehalt *m*

1858 moisture content of snow
f teneur *f* en eau de la neige
e humedad *f* de la nieve
i umidità *f* della neve
p umidade *f* da neve
d Schneefeuchte *f*

1859 moisture deficiency
f manque *m* d'humidité
e deficiencia *f* de humedad
i deficienza *f* d'umidità
p deficiência *f* de umidade
d Feuchtigkeitsdefizit *n*

1860 moisture potential
f potentiel *m* hydrique
e potencial *m* de humedad
i potenziale *m* idrico
p potencial *m* de água
d Feuchtigkeitspotential *n*

1861 moisture-retentive *adj*
f retenant l'humidité
e hidrófilo
i che conserva l'umidità
p retentor da umidade
d feuchthaltig

1862 moisture trap
f piège *f* à humidité
e trampa *f* de humedad
i trappola *f* d'umidità
p recuperador *m* de umidade
d Kühlfalle *f*

1863 moist warm air
f air *m* humide et chaud
e aire *m* húmedo y caliente
i aria *f* umida e calda
p ar *m* úmido e quente
d feuchtwarme Luft *f*

1864 molecular oxygen
f oxygène *m* moléculaire
e oxígeno *m* molecular
i ossigeno *m* molecolare
p oxigênio *m* molecular
d molekularer Sauerstoff *m*

1865 moment
f moment *m*
e momento *m*
i momento *m*
p momento *m*
d Moment *m*

1866 monsoon
(seasonal wind from Indian Ocean to Japan and western Pacific blowing from the southwest from April to October, and from the northeast during the remainder of the year influencing the climate of large regions)
f mousson *f*
e monzón *m*
i monsone *m*
p monção *f*
d Monsun *m*

1867 monsoon climate
f climat *m* de mousson
e clima *m* monzónico
i clima *m* monsonico
p clima *m* de monções
d Monsunklima *n*

1868 monsoon drift
f courant *m* des moussons
e corriente *f* de los monzones
i corrente *f* dei monsoni
p corrente *f* das monções
d Monsundrift *f*

1869 month
f mois *m*
e mes *m*
i mese *m*
p mês *m*
d Monat *m*

1870 moon
f lune *f*
e luna *f*
i luna *f*
p lua *f*
d Mond *m*

*** moon dog → 2017**

1871 moonless *adj*
f sans lune
e sin luna
i senza luna
p sem lua
d mondlos

1872 moon quarter; quarter of moon
f quartier *m* de la lune
e cuarto *m* de luna
i quarto *m* di luna
p quarto *m* de lua
d Mondviertel *n*

1873 moraine
f moraine *f*
e morrena *f*
i morena *f*
p morena *f*
d Moräne *f*

1874 morning; forenoon
f matin *m*
e mañana *f*
i mattino *m*
p manhã *f*
d Vormittag *m*

*** morning, in the ~ → 1521**

*** morning twilight → 723**

1875 morphological types of glaciers
f types *mpl* morphologiques des glaciers
e tipos *mpl* morfológicos de los glaciares
i tipi *mpl* morfologici del ghiacciai
p tipos *mpl* morfológicos das geleiras
d morphologische Gletschertypen *fpl*

* **mottled umber moth → 2402**

1876 mould *adj*
f moisi
e enmohecido
i ammuffito
p molde
d schimmelig

1877 mountain
f montagne *f*
e montaña *f*
i montagna *f*
p montanha *f*
d Gebirge *n*

1878 mountain barometer
f baromètre *m* de montagne
e barómetro *m* de montaña
i barometro *m* di montagna
p barômetro *m* de montanha
d Höhenbarometer *n*

1879 mountain breeze
f brise *f* de montagne
e brisa *f* de montaña
i brezza *f* di montagna
p brisa *f* da montanha
d Bergwind *m*

1880 mountain climate
f climat *m* de montagne
e clima *m* de montaña
i clima *m* di montagna
p clima *m* de montanha
d Höhenklima *n*

1881 mountaineering
f alpinisme *m*
e alpinismo *m*
i alpinismo *m*
p alpinismo *m*
d Bergsteigen *n*

1882 mountain entrapment
f captation *f* des eaux de montagne
e embalse *m* de las aguas rodadas
i allacciamento *m* montano
p captação *f* das águas de montanha
d Talsperre *f*

1883 mountain glaciation
f glaciation *f* de montagne
e glaciación *f* de montaña
i glaciazione *f* di montagna
p glaciação *f* de montanha
d Gebirgsvergletscherung *f*

1884 mountain glacier
f glacier *m* de montagne
e glaciar *m* alpino
i ghiaccio *m* alpino
p glaciar *m* alpino
d Berggletscher *m*

1885 mountain sickness
f mal *m* d'altitude
e mal *m* de la altitud
i malattia *f* di montagna
p mal-de-montanha *m*
d Höhenkrankheit *f*

1886 mountain torrent
f eau *f* impétueuse
e agua *f* torrencial
i acqua *f* irruente
p água *f* torrencial
d Wildbach *m*

1887 mud
f vase *f*
e fango *m*
i fango *m*
p lama *f*
d Schlamm *m*

1888 mud bank
f banc *m* de vase
e banco *m* de fango
i banco *m* di fango
p banco *m* de lama
d Schlammbank *f*

1889 muddy *adj*
f vaseux
e fangoso
i fangoso
p lamacento
d schlammig

1890 muddy water
f eau *f* vaseuse
e agua *f* fangosa
i acqua *f* fangosa
p água *f* lodosa
d Schlickwasser *n*

1891 mud fever
f fièvre *f* des eaux
e fiebre *f* de las aguas

i febbre *f* dell'acque
p febre *f* das águas
d Feldfieber *n* A *(durch Leptospira grippotyphosa)*

1892 mudflow
f avalanche *f* boueuse
e alud *m* de lama
i torrente *m* di fango
p avalancha *f* de lama
d Schlammstrom *m*

* **muggy weather → 2619**

1893 mull
f humus *m* doux
e mantillo *m* suave
i humus *m* saturo
p mull *m*
d Mull *m*

1894 mutual neutralization
f neutralisation *f* mutuelle
e neutralización *f* mutua
i neutralizzazione *f* mutua
p neutralização *f* mútua
d gegenseitige Neutralisation *f*

N

* **national park** → 1903

1895 natural calamity
f calamité *f* naturelle
e catástrofe *f* natural
i calamità *f* naturale
p catástrofe *f* natural
d Naturkatastrophe *f*

1896 natural convection
f convection *f* naturelle
e convección *f* natural
i convezione *f* naturale; convezione *f* spontanea
p convecção *f* natural
d natürliche Konvektion *f*

1897 natural ice
f glace *f* naturelle
e hielo *m* natural
i ghiaccio *m* naturale
p gelo *m* natural
d natürliches Eis *n*

1898 natural radioactivity
f radioactivité *f* naturelle
e radiactividad *f* natural
i radioattività *f* naturale
p radioatividade *f* natural
d natürliche Radioaktivität *f*

1899 natural resources
f ressources *fpl* naturelles
e recursos *mpl* naturales
i risorse *fpl* naturali
p recursos *mpl* naturais
d natürliche Hilfsquellen *fpl*

1900 natural water content
f teneur *f* en eau naturelle
e humedad *f* natural
i contenuto *m* naturale d'acqua
p teor *m* de umidade natural
d natürlicher Wassergehalt *m*

1901 nature management; protection of nature; conservation of nature
f gestion *f* de l'environnement; protection *f* des sites
e mantenimiento *m* de la naturaleza
i gestione *f* della natura; conservazione *f* della natura
p proteção *f* da natureza
d Naturverwaltung *f*; Naturschutz *m*

1902 nature of the soil
f nature *m* du sol
e naturaleza *f* del suelo
i natura *f* del suolo
p natureza *f* do solo
d Bodenbeschaffenheit *f*

1903 nature reserve; national park
f réserve *f* naturelle; parc *m* national
e reserva *f* natural
i parco *m* naturale; riserva *f* naturale
p reserva *f* natural
d Naturschutzgebiet *n*

1904 nautical mile
f mille *m* marin
e milla *f* marina
i miglio *m* nautico
p milha *f* náutica
d Seemeile *f*

1905 nautical twilight
f crépuscule *m* nautique
e crepúsculo *m* náutico
i crepuscolo *m* nautico
p crepúsculo *m* náutico
d nautische Dämmerung *f*

1906 neap tide
f marée *f* d'eau morte
e marea *f* muerta
i marea *f* alle quadrature
p maré *f* de águas mortas
d Nipptide *f*

* **nebulosity** → 550

1907 neon
f néon *m*
e neón *m*
i neo *m*
p neônio *m*
d Neon *n*

1908 nephelometer
f néphélémètre *m*
e nefelémetro *m*
i nefelometro *m*
p nefelômetro *m*
d Nephelometer *n*

1909 nephology
f néphologie *f*
e nefología *f*

i nefologia *f*; studio *m* delle nubi
p nefologia *f*
d Wolkenkunde *f*

1910 nephometric scale
f échelle *f* néphométrique
e escala *f* nefométrica
i scala *f* nefometrica
p escala *f* nefométrica
d nephometrischer Maßstab *m*

1911 nephoscope
f néphoscope *m*
e nefoscopio *m*
i nefoscopio *m*
p nefoscópio *m*
d Wolkenspiegel *m*

1912 net precipitation
f précipitation *f* nette
e precipitación *f* neta
i precipitazione *f* netta
p precipitação *f* líquida
d Nettoniederschlag *m*

1913 network
f réseau *m*
e red *f*
i rete *f*
p rede *f*
d Netzwerk *n*

1914 neutral point
f point *m* neutre
e punto *m* neutro
i punto *m* neutro
p ponto *m* neutro
d Neutralpunkt *m*

1915 neutrosphere
f neutrosphère *f*
e neutrosfera *f*
i neutrosfera *f*
p neutrosfera *f*
d Neutrosphäre *f*

* **new ice → 3084**

1916 new moon
f nouvelle lune *f*
e luna *f* nueva
i luna *f* nuova
p lua *f* nova
d Neumond *m*

1917 night
f nuit *f*
e noche *f*
i notte *f*
p noite *f*
d Nacht *f*

1918 night airglow; night glow
f lueur *f* atmosphérique nocturne
e resplandor *m* de aire nocturno
i chiarore *m* notturno dell'atmosfera
p luminescência *f* do céu noturno
d Nachthimmelsleuchten *n*

* **night, at ~ → 266**

1919 night cooling
f refroidissement *m* nocturne
e enfriamiento *m* nocturno
i raffreddamento *m* notturno
p arrefecimento *m* noturno
d Nachtabkühlung *f*

* **nightfall, at ~ → 267**

1920 night frost
f gelée *f* nocturne
e helada *f* de noche; helada *f* nocturna
i gelata *f* notturna
p geada *f* noturna
d Nachtfrost *m*

* **night glow → 1918**

1921 night interruption
f interruption *f* de la nuit
e interrupción *f* de la noche
i interruzione *f* notturna
p interrupção *f* da noite
d Nachtunterbrechung *f*

1922 night sky
f ciel *m* nocturne
e cielo *m* nocturno
i cielo *m* notturno
p céu *m* noturno
d Nachthimmel *m*

1923 nimbostratus
(type of cloud characterized by diffuse dark-grey appearance)
f nimbo-stratus *m*
e nimbostratus *m*
i stratonembo *m*
p nimbostratus *m*
d Nimbostratus *m*

1924 nimbus
(type of cloud characterized by bringing rain or snow)
f nimbus *m*
e nimbo *m*

i nembo *m*
p nimbo *m*; nimbus *m*
d Nimbuswolke *f*

1925 nitrogen
f azote *m*
e nitrógeno *m*
i azoto *m*
p nitrogênio *m*
d Nitrogen *n*

1926 nitrogen dioxide
f bioxyde *m* d'azote
e dióxido *m* de nitrógeno
i biossido *m* di azoto
p dióxido *m* de nitrogênio
d Stickstoffdioxid *n*

1927 nival zone
f zone *f* nivale
e zona *f* niveal
i zona *f* nivale
p zona *f* niveal
d Schneezone *f*

* **noble gas → 1499**

1928 noctilucent clouds
f nuages *mpl* noctilucents
e nubes *fpl* noctilucentes
i nubi *fpl* nottilucenti
p nuvens *fpl* noctilucentes
d nachtleuchtende Wolken *fpl*

1929 nomogram
f nomogramme *m*
e nomograma *m*
i nomogramma *m*
p nomograma *m*
d Nomogramm *n*

1930 nomography
f nomographie *f*
e nomografía *f*
i nomografia *f*
p nomografia *f*
d Nomographie *f*

1931 normal barometer
f baromètre *m* normal
e barómetro *m* normal
i barometro *m* normale
p barômetro *m* normal
d Normalbarometer *n*

1932 normal distribution
f distribution *f* normale
e distribución *f* normal
i distribuzione *f* normale
p distribuição *f* normal
d Gaussche Fehlerverteilung *f*

1933 normal sound
f son *m* normal
e sonido *m* normal
i suono *m* normale
p som *m* normal
d normaler Schall *m*

1934 north
f nord *m*
e norte *m*
i nord *m*
p norte *m*
d Norden *m*

1935 northeast monsoon
f mousson *f* du nord-est
e monzón *m* del nordeste
i monsone *m* da nord-est
p monção *f* de nordeste
d Nordostmonsum *m*

1936 northeast trade wind
f alizé *m* du nord-est
e alisio *m* del nordeste
i aliseo *m* da nord-est
p alísio *m* de nordeste
d Nordostpassat *m*

1937 northeast wind
f vent *m* du nord-est
e viento *m* del nordeste
i vento *m* da nord-est
p vento *m* de nordeste
d Nordostwind *m*

1938 north equatorial current
f courant *m* équatorial nord
e corriente *f* ecuatorial norte
i corrente *f* equatoriale da nord
p corrente *f* equatorial de norte
d Nordäquatorialstrom *m*

1939 norther
f norther *m*; vent *m* nord
e nortada *f*
i vento *m* da nord
p vento *m* frio de norte; aquilão *m*
d Norder *m*; Nordsturm *m*

1940 northern celestial hemisphere
f hémisphère *m* céleste boréal
e hemisferio *m* celeste boreal
i emisfero *m* celeste boreale
p hemisfério *m* celeste boreal
d nördliche Himmelshalbkugel *f*

1941 northern hemisphere
f hémisphère *m* boréal
e hemisferio *m* boreal
i emisfero *m* boreale
p hemisfério *m* boreal
d nördliche Halbkugel *f*

* **north polar circle → 205**

1942 north pole
f pôle *m* nord
e polo *m* norte
i polo *m* nord
p pólo *m* norte
d Nordpol *m*

1943 North Sea
f Mer *f* du Nord
e Mar *m* del Norte
i Mare *m* da Nord
p Mar *m* do Norte
d Nordsee *f*

1944 northwest
f nord-ouest *m*
e noroeste *m*
i nord-ovest *m*
p noroeste *m*
d Nordwesten *m*

1945 northwest monsoon
f mousson *f* du nord-ouest
e monzón *m* del noroeste
i monsone *m* da nord-ovest
p monção *f* de noroeste
d Nordwestmonsun *m*

1946 northwest wind
f vent *m* du nord-ouest
e viento *m* del noroeste
i vento *m* da nord-ovest
p vento *m* de noroeste
d Nordwestwind *m*

1947 north wind
f vent *m* du nord
e viento *m* del norte
i vento *m* da nord
p vento *m* de norte
d Nordwind *m*

1948 November
f novembre *m*
e noviembre *m*
i novembre *m*
p novembro *m*
d November *m*

1949 nuclear winter
f hiver *m* nucléaire
e invierno *m* nuclear
i inverno *m* nucleare
p inverno *m* nuclear
d Kernwinter *m*

1950 nucleation
f nucléation *f*
e nucleación *f*; formación *f* de núcleos de condensación
i nucleazione *f*
p nucleação *f*; semeadura *f* de nuvens
d Nukleation *f*; Kernbildung *f*

1951 number of particles
f nombre *m* des particules
e número *m* de partículas
i numero *m* delle particelle
p número *m* de partículas
d Partikelzahl *f*; Zahl der Teilchen *f*

O

1952 oasis
f oasis *f*
e oasis *m*
i oasi *f*
p oásis *m*
d Oase *f*

1953 oblateness of the earth; earth's oblateness
f aplatissement *m* de la terre
e achatamiento *m* de la tierra
i schiacciamento *m* terrestre
p achatamento *m* da terra
d Erdabplattung *f*

1954 obliquity of the ecliptic
f obliquité *f* de l'écliptique
e oblicuidad *f* de la elíptica
i obliquità *f* dell'eclittica
p obliquidade *f* da eclíptica
d Schiefe *f* der Ekliptik

1955 observer
f observateur *m*
e observador *m*
i osservatore *m*
p observador *m*
d Beobachter *m*

1956 occasional *adj*
f occasionnel
e ocasional
i occasionale
p ocasional
d gelegentlich

*** occluded front → 1957**

1957 occlusion; occluded front
f occlusion *f*
e oclusión *f*; frente *m* ocluido
i occlusione *f*; fronte *m* occluso
p oclusão *f*; frente *f* absorvida
d Okklusion *f*; Okklusionsfront *f*

1958 ocean
f océan *m*
e océano *m*
i oceano *m*
p oceano *m*
d Ozean *m*

1959 ocean current; stream current
f courant *m* océanique
e corriente *f* oceánica
i corrente *f* oceanica
p corrente *f* oceânica
d ozeanische Strömung *f*

1960 ocean glaciology
f glaciologie *f* des océans
e glaciología *f* de los océanos
i glaciologia *f* degli oceani
p glaciologia *f* dos oceanos
d Glaziologie *f* der Meere

1961 oceanography
f océanographie *f*
e oceanografía *f*
i oceanografia *f*
p oceanografia *f*
d Ozeanographie *f*

1962 octant
f octant *m*
e octante *m*
i ottante *m*
p octante *m*
d Oktant *m*

1963 October
f octobre *m*
e octubre *m*
i ottobre *m*
p outubro *m*
d Oktober *m*

1964 oil-contaminated waters
f eaux *fpl* contaminées par l'huile
e aguas *fpl* contaminadas por aceite
i acque *fpl* contaminate per olio
p águas *fpl* contaminadas por óleo
d ölverseuchte Gewässer *n*

1965 oil contamination
f contamination *f* par l'huile
e contaminación *f* por aceite
i contaminazione *f* per olio
p contaminação *f* por óleo
d Ölverseuchung *f*

1966 old ice
f glace *f* vieille
e hielo *m* viejo
i ghiaccio *m* vecchio
p gelo *m* velho
d Alteis *n*

1967 old snow
f vieille neige *f*
e nieve *f* vieja

i neve *f* vecchia
p neve *f* velha
d Altschnee *m*

1968 on account of bad weather
f par suite de mauvais temps
e a causa de mal tiempo
i in conseguenza di cattivo tempo
p em consequência do mau tempo
d wegen schlechten Wetters

1969 on-shore wind
(breeze blowing from the sea toward the land)
f vent *m* du large
e viento *m* hacia tierra
i vento *m* di mare
p vento *m* do mar
d auflandiger Wind *m*

1970 on-the-floor drying
f séchage *m* sur le sol
e secado *m* en el suelo
i essiccazione *f* al suolo
p secado *m* ao solo
d Bodentrocknung *f*

1971 opalescence
f opalescence *f*
e opalescencia *f*
i opalescenza *f*
p opalescência *f*
d Opaleszenz *f*

1972 open water; water free of ice
f mer *f* libre; mer *f* libre de glaces
e mar *m* libre; mar *m* libre de hielo
i mare *m* libero; mare *m* libero di ghiacci
p mar *m* livre; mar *m* livre de gelo
d offenes Wasser *n*; eisfreies Wasser *n*

* **oppressive weather → 2619**

1973 optical phenomenon
f phénomène *m* optique
e fenómeno *m* óptico
i fenomeno *m* ottico
p fenômeno *m* ótico
d Lichterscheinung *f*

1974 ordinary wave
f onde *f* ordinaire; rayon *m* ordinaire
e onda *f* ordinaria
i onda *f* ordinaria
p onda *f* ordinária
d ordentliche Welle *f*

1975 original heat
f chaleur *f* originelle
e calor *m* natural
i calore *m* originario
p calor *m* natural
d ursprüngliche Wärme *f*

1976 original soil
f sol *m* en place
e suelo *m* natural
i terreno *m* vergine
p solo *m* natural
d gewachsener Boden *m*

1977 orographic cloud
f nuage *m* orographique
e nube *f* orográfica
i nube *f* orografica
p nuvem *f* orográfica
d Stauwolke *f*

1978 orographic rain
f pluie *f* orographique
e lluvia *f* orográfica
i pioggia *f* orografica
p chuva *f* orográfica
d Stauregen *m*

1979 orographic snow line
f limite *f* orographique de la neige
e línea *f* orográfica de la nieve
i linea *f* orografica della neve
p linha *f* orográfica de la neve
d orographische Schneegrenze *f*

1980 orography
(the branch of geography dealing with mountains and mountain systems)
f orographie *f*
e orografía *f*
i orografia *f*
p orografia *f*
d Orographie *f*

1981 oscillation
f oscillation *f*
e oscilación *f*
i oscillazione *f*
p oscilação *f*
d Oszillation *f*

1982 osmometer
f osmomètre *m*
e osmómetro *m*
i osmometro *m*
p osmômetro *m*
d Osmometer *n*

1983 outdoor crop; field crop
f culture *f* en plein air
e cultivo *m* al aire libre
i coltura *f* all'aperto; coltura *f* in pieno campo

p cultura *f* de ar livre
d Freilandkultur *f*

1984 outlet glacier
f émissaire *m* glaciaire
e glaciar *m* de terminal
i ghiacciaio *m* terminale
p geleira *f* de descarga
d Gletscherstrom *m*

1985 outside air
f air *m* extérieur
e aire *m* exterior
i aria *f* esterna
p ar *m* exterior
d Außenluft *f*

1986 outside-air cooling
f refroidissement *m* par l'air froid pris à l'extérieur
e enfriamiento *m* por aire externo
i raffreddamento *m* con aria esterna
p arrefecimento *m* por ar frio do exterior
d Außenluftkühlung *f*

1987 overcast *adj*
f couvert
e cubierto
i coperto
p encoberto
d bedeckt

1988 overcast day
f jour *m* couvert
e día *m* cubierto
i giorno *m* caliginoso
p dia *m* enevoado
d trüber Tag *m*

* **overcast sky → 548**

* **overflow → 1523**

1989 overflow weir
f déversoir *m* en chute libre
e desviador *m* del agua de tormenta
i sfioratore *m* a libera caduta
p escoadouro *m* da água da tormenta
d Überfallwehr *n*

1990 oxygen
f oxygène *m*
e oxígeno *m*
i ossigeno *m*
p oxigênio *m*
d Sauerstoff *m*

1991 oxygen absorption
f absorption *f* d'oxygène
e absorción *f* de oxígeno
i assorbimento *m* di ossigeno
p absorção *f* de oxigênio
d Sauerstoffaufnahme *f*

1992 oxygen apparatus
f appareil *m* d'oxygène
e aparato *m* de oxígeno
i apparecchio *m* di ossigeno
p aparelho *m* de oxigênio
d Sauerstoffapparat *m*

1993 oxygen saturation
f saturation *f* de oxygène
e saturación *f* de oxígeno
i saturazione *f* di ossigeno
p saturação *f* de oxigênio
d Sauerstoffsättigung *f*

1994 ozone
f ozone *m*
e ozono *m*
i ozono *m*
p ozônio *m*
d Ozon *m*

* **ozone layer → 1997**

1995 ozone shadow
f ombre *f* d'ozone; absorption *f* par ozone
e sombra *f* del ozono
i ombra *f* dell'ozono
p sombra *f* do ozônio
d Ozonschatten *m*

1996 ozonized air
f air *m* ozonisé
e aire *m* ozonizado
i aria *f* ozonizzata
p ar *m* ozonizado
d ozonhaltige Luft *f*

1997 ozonosphere; ozone layer
f ozonosphère *f*; couche *f* d'ozone
e ozonosfera *f*
i ozonosfera *f*; strato *m* di ozono
p ozonosfera *f*; camada *f* de ozônio
d Ozonosphäre *f*; Ozonschicht *f*

P

1998 Pacific Ocean
f Océan *m* pacifique
e Océano *m* pacífico
i Oceano *m* pacifico
p Oceano *m* pacífico
d Stiller Ozean *m*

1999 pack ice
f glas *m* de dérive
e banco *m* de hielos a la deriva
i ghiaccio *m* ammonticchiato
p pacote *m* de gelo
d Treibeis *n*

2000 pad-and-fan cooling
f refroidissement *m* par ventilation et pad
e enfriamiento *m* con evaporador y ventilador
i raffreddamento *m* con pannelli e ventilatori
p arrefecimento *m* forçado com evaporação de água
d Mattenkühlung *f*

2001 pal(a)eoatmosphere
f paléoatmosphère *f*
e paleoatmósfera *f*
i paleoatmosfera *f*
p paleoatmosfera *f*
d Paläoatmosphäre *f*

2002 pal(a)eoclimate
f paléoclimat *m*
e paleoclima *m*
i paleoclima *m*
p paleoclima *m*
d Paläoklima *n*

2003 pal(a)eoclimatology
f paléoclimatologie *f*
e paleoclimatología *f*
i paleoclimatologia *f*
p paleoclimatologia *f*
d Paläoklimatologie *f*

2004 pal(a)eoecology
f paléoécologie *f*
e paleoecología *f*
i paleoecologia *f*
p paleoecologia *f*
d Paläoökologie *f*

2005 pal(a)eoglaciology
f paléoglaciologie *f*
e paleoglaciología *f*
i paleoglaciologia *f*
p paleoglaciologia *f*
d Paläoglaziologie *f*

2006 pal(a)eomagnetism
f paléomagnétisme *m*
e paleomagnetismo *m*
i paleomagnetismo *m*
p paleomagnetismo *m*
d Paläomagnetismus *m*

2007 pal(a)eotemperature analysis
f analyse *f* paléothermique
e análisis *f* paleotérmica
i analisi *f* paleotermica
p análise *f* paleotérmica
d Analyse *f* der Paläotemperaturen

2008 pallium
(extended sheet of clouds)
f pallium *m*
e palio *m*
i pallium *m*
p pallium *m*
d Pallium *n*

2009 paludification
f transformation *f* en marais
e transformación *f* en pantano
i impaludamento *m*; trasformazione *f* in palude
p transformação *f* em pântano
d Versumpfen *n*

2010 pampa
(extensive plain covered with underbrush, in South America mainly: Argentina, Uruguay and Rio Grande do Sul (Brazil))
f pampa *f*
e pampa *f*
i pampa *f*
p pampa *f*
d Pampa *f*

2011 pampero
(strong wind blowing from the southwest across the pampas of Argentina, Uruguay and Rio Grande do Sul (Brazil))
f pampéro *m*
e pampero *m*
i pampero *m*
p pampero *m*
d Pampero *m*

2012 pancake ice
f glace *f* en crêpes
e hielo *m* panqueque
i ghiaccio *m* a frittelle
p gelo-panqueca *m*
d Pfannkucheneis *n*

2013 parallax
f parallaxe *f*
e paralaje *f*
i parallasse *f*
p paralaxe *f*
d Parallaxe *f*

2014 parallel
f parallèle *m*
e paralelo *m*
i parallelo *m*
p paralelo *m*
d Parallele *f*

2015 parameter
f paramètre *m*
e parámetro *m*
i parametro *m*
p parâmetro *m*
d Parameter *m*

2016 paranthelion
f paranthélie *m*
e parantelio *m*
i parantelio *m*
p parantélio *m*
d Nebengegensonne *f*

2017 paraselene; moon dog
f parasélène *f*
e paraselena *f*
i paraselene *f*
p parasselênio *m*
d Nebenmond *m*

2018 parhelic circle
f cercle *m* parhélique
e círculo *m* parhélico
i circolo *m* parelico
p círculo *m* parélico
d Sonnenhof *m*

2019 parhelion; mock sun
f parhélie *m*
e parhelio *m*
i parelio *m*
p parélio *m*
d Nebensonne *f*

2020 partial eclipse of the moon
f éclipse *f* partielle de la lune
e eclipse *m* parcial de la luna
i eclisse *f* parziale di luna
p eclipse *m* parcial da lua
d partielle Mondfinsternis *f*

2021 particulates
f particules *fpl*
e partículas *fpl*
i polveri *fpl*
p partículas *fpl*
d Makroteilchen *npl*

* **partly cloudy** *adj* → **1016**

2022 pascal
(unit of pressure in the meter-kilogram-second international system, due to the pressure resulting from a force of 1 newton per square meter)
f pascal *m*
e pascal *m*
i pascal *m*
p pascal *m*
d Pascal *n*

2023 passing shower
f averse *f* passagère
e chubasco *m* pasajero
i rovescio *m*; pioggia *f* di breve durata
p chuvarada *f*; aguaceiro *m* passageiro
d Schauer *m*

2024 pattern
f configuration *f*
e modelo *m*
i procedura *f*
p configuração *f*
d Modell *n*

2025 peak
f maximum *m* de la région
e pico *m*; cerro *m*
i massimo *m*; livello *m* di massimo
p pico *m*
d Gipfel *m*; Maximum *n* einer Schicht

* **pedogenesis** → **2448**

* **pedology** → **2462**

2026 pedosphere
f pédosphère *f*
e pedosfera *f*
i pedosfera *f*
p pedosfera *f*
d Pedosphäre *f*

2027 pendulum anemometer
f anémomètre *m* à pendule
e anemómetro *m* de péndulo

i anemometro *m* a pendolo
p anemômetro *m* de pêndulo
d Pendelwindmesser *m*

2028 peneplain
f pénéplaine *f*
e penillanura *f*
i piano *m* di livello di base
p peneplano *m*
d Peneplain *f*

2029 pennant
f fanion *m*
e banderola *f*
i bandierina *f*
p flâmula *f*
d Wimpel *m*

2030 penumbra
f pénombre *f*
e penumbra *f*
i penombra *f*
p penumbra *f*
d Halbschatten *m*

2031 penumbra of the earth
f pénombre *f* de la terre
e penumbra *f* de la tierra
i penombra *f* della terra
p penumbra *f* da terra
d Halbschatten *m* der Erde

2032 perched water
f accumulation *f* d'eau
e contención *f* del agua
i ristagno *m* d'acqua
p contenção *f* da água
d Wasserstau *m* im Boden

2033 perched water table
f humidité *f* de stagnation
e humedad *f* del agua gravitante
i superficie *f* piezometrica della falda sospesa
p umidade *f* estagnada
d Staunässe *f*

2034 perfectly dry air
f air *m* absolument sec
e aire *m* completamente seco
i aria *f* completamente secca
p ar *m* absolutamente seco
d vollkommen trockene Luft *f*

2035 perigee
f périgée *m*
e perigeo *m*
i perigeo *m*
p perigeu *m*
d Erdnähe *f*

2036 periglacial *adj*
f périglaciaire
e periglacial
i periglaciale
p periglacial
d periglazial

2037 perihelion
f périhélie *m*
e perihelio *m*
i perielio *m*
p periélio *m*
d Perihelium *n*

2038 periodical wind
f vent *m* périodique
e viento *m* periódico
i vento *m* periodico
p vento *m* periódico
d periodischer Wind *m*

2039 permafrost
(permanent cooling of ground, subsoil and rocks in arctic regions)
f pergélisol
e terreno *m* helado permanentemente
i permagelo *m*
p subsolo *m* permanentemente gelado
d Dauerfrostboden *m*; Permafrost *m*

2040 permanent magnetism
f magnétisme *m* permanent
e magnetismo *m* permanente
i magnetismo *m* permanente
p magnetismo *m* permanente
d bleibender Magnetismus *m*

2041 permeability
f perméabilité *f*
e permeabilidad *f*
i permeabilità *f*
p permeabilidade *f*
d Permeabilität *f*

2042 perpetual frost climate
f climat *m* de gel perpétuel
e clima *m* del hielo eterno
i clima *m* a gelo perpetuo
p clima *m* do gelo eterno
d Klima *n* ewigen Frostes

2043 personal equation
f équation *f* personnelle
e ecuación *f* personal
i equazione *f* personale
p equação *f* pessoal
d persönliche Gleichung *f*

2044 perturbance
f perturbation *f*
e perturbación *f*
i perturbazione *f*
p perturbação *f*
d Störung *f*

2045 Peru Current; Humboldt Current
f Courant *m* de Pérou; Courant *m* de Humboldt
e Corriente *f* del Perú; Corriente *f* de Humboldt
i Corrente *f* del Peru; Corrente *f* Humboldt
p Corrente *f* do Peru; Corrente *f* de Humboldt
d Perustrom *m*; Humboldtstrom *m*

2046 petrography of ice
f pétrographie *f* de la glace
e petrografía *f* del hielo
i petrografia *f* del ghiaccio
p petrografia *f* do gelo
d Eispetrographie *f*

2047 phase of the moon
f phase *f* de la lune
e fase *f* lunar
i fase *f* lunare
p fase *f* lunar
d Mondphase *f*

2048 phenomenon
f phénomène *m*
e fenómeno *m*
i fenomeno *m*
p fenômeno *m*
d Phänomen *n*

2049 photology
f photologie *f*
e fotología *f*
i fotologia *f*
p fotologia *f*
d Photologie *f*; Lehre *f* vom Licht

2050 photoluminescence
f photoluminescence *f*
e fotoluminiscencia *f*
i fotoluminescenza *f*
p fotoluminescência *f*
d Photolumineszenz *f*

2051 photometer
f photomètre *m*
e fotómetro *m*
i fotometro *m*
p fotômetro *m*
d Photometer *n*

2052 photometry
f photométrie *f*
e fotometría *f*
i fotometria *f*
p fotometria *f*
d Photometrie *f*

2053 photoperiodism
f photopériodisme *m*
e fotoperiodismo *m*
i fotoperiodismo *m*
p fotoperiodismo *m*
d Photoperiodismus *m*

2054 photosphere
f photosphère *f*
e fotosfera *f*
i fotosfera *f*
p fotosfera *f*
d Photosphäre *f*

2055 photosynthesis
f photosynthèse *f*
e fotosíntesis *f*
i fotosintesi *f*
p fotosíntese *f*
d Photosynthese *f*

2056 phototropism
f phototropisme *m*
e fototropismo *m*
i fototropismo *m*
p fototropismo *m*
d Phototropismus *m*

*** phreatic water → 1279**

2057 physics of ice
f physique *m* de la glace
e física *f* del hielo
i fisica *f* del ghiaccio
p física *f* do gelo
d Eisphysik *f*

2058 physiography
f physiographie *f*
e fisiografía *f*
i fisiografia *f*
p fisiografia *f*
d physikalische Geographie *f*; Physiogeographie *f*

2059 physiological atmospheric humidity
f humidité *f* atmosphérique physiologique
e humedad *f* atmosférica fisiológica
i umidità *f* atmosferica fisiologica
p umidade *f* atmosférica fisiológica
d physiologische Luftfeuchtigkeit *f*

2060 physiology
f physiologie *f*
e fisiología *f*
i fisiologia *f*
p fisiologia *f*
d Physiologie *f*

2061 phytoplankton
f phytoplancton *m*
e fitoplancton *m*
i fitoplancton *m*
p fitoplâncton *m*
d Phytoplankton *n*

2062 phytotron
f phytotron *m*
e fitotrón *m*
i fitotrone *m*
p fitotron *m*
d Phytotron *n*

2063 picking season
f époque *f* de la cueillette
e época *f* de recolección
i stagione *f* della raccolta
p época *f* de colheita
d Erntezeit *f*

2064 piezometer
f piézomètre *m*
e piezómetro *m*
i piezometro *m*
p piezômetro *m*
d Piezometer *n*

2065 piezometric level
f niveau *m* piézométrique
e nivel *m* piezométrico
i quota *f* piezometrica
p nível *m* piezométrico
d Piezometerspiegel *m*

2066 pileus
(cloud resembling a cap or scarf)
f pileus *m*
e pileus *m*
i pileus *m*
p pileus *m*
d Piläuswolke *f*

2067 pilot balloon
f ballon-pilote *m*
e globo *m* piloto
i pallone *m* pilota
p balão *m* piloto
d Pilotballon *m*

2068 Pitot tube
f tube *m* de Pitot
e tubo *m* de Pitot
i tubo *m* di Pitot
p tubo *m* de Pitot
d Pitotröhre *f*

2069 Planck's law
f loi *f* de Planck
e ley *f* de Planck
i legge *f* di Planck
p lei *f* de Planck
d Plancksches Gesetz *n*

2070 planetary motion
f mouvement *m* planétaire
e movimiento *m* planetario
i movimento *m* planetario
p movimento *m* planetário
d Planetenbewegung *f*

2071 plankton
f plancton *m*
e plancton *m*
i plancton *m*
p plâncton *m*
d Plankton *n*

*** plant cover → 2890**

2072 planting season
f époque *f* de plantation
e época *f* de plantación
i stagione *f* di semina
p época *f* de plantação
d Pflanzzeit *f*

2073 plant kingdom
f règne *m* végétal
e reino *m* vegetal
i regno *m* vegetable
p reino *m* vegetal
d Pflanzenreich *n*

2074 plant protection; crop protection
f protection *f* des plantes
e protección *f* de las plantas
i difesa *f* delle piante
p proteção *f* das plantas
d Pflanzenschutz *m*

2075 plasmolysis
f plasmolyse *f*
e plasmólisis *f*
i plasmolisi *f*
p plasmólisis *f*
d Plasmolyse *f*

2076 plastic cladding
f revêtement *m* de plastique
e revestimiento *m* de plástico

i rivestimento *m* in plastica
p cobertura *f* de plástico
d Folienauskleidung *f*

2077 plastic greenhouse
f abri-serre *f* de plastique
e invernadero *m* de lámina de plástico
i serra *f* ricoperta in film di plastica
p estufa *f* de plástico
d Folienhaus *n*

2078 plasticity of ice
f plasticité *f* de la glace
e plasticidad *f* del hielo
i plasticità *f* del ghiaccio
p plasticidade *f* do gelo
d Eisplastizität *f*

2079 plasticity of snow
f plasticité *f* de la neige
e plasticidad *f* de la nieve
i plasticità *f* della neve
p plasticidade *f* da neve
d Schneeplastizität *f*

* **plateau glacier → 629**

2080 plate ice
f glace *f* en plaques
e hielo *m* en planchas
i ghiaccio *m* in lastre
p gelo *m* em pranchas
d Platteneis *n*

2081 pluviograph; rainfall recorder; recording rain gauge
f pluviographe *m*; pluviomètre *m*
e pluviógrafo *m*; registro *m* pluviométrico
i pluviografo *m*; pluviometro *m*
p pluviógrafo *m*; registro *m* pluviométrico
d Pluviograph *m*; Regenschreiber *m*

* **p.m. → 62**

2082 pocket barometer
f baromètre *m* de poche
e barómetro *m* de bolsillo
i barometro *m* tascabile
p barômetro *m* de algibeira
d Taschenbarometer *n*

2083 Poisson's ratio of ice
f coefficient *m* de Poisson de la glace
e coeficiente *m* de Poisson del hielo
i coefficiente *m* di Poisson del ghiaccio
p coeficiente *m* de Poisson do gelo
d Poissonzahl *f*; Kontraktionskoeffizient *m* des Eises

2084 polar *adj*
f polaire
e polar
i polare
p polar
d polar

2085 polar angle
f angle *m* polaire
e ángulo *m* polar
i angolo *m* polare
p ângulo *m* polar
d Polarwinkel *m*

2086 polar axis
f axe *m* polaire
e eje *m* polar
i asse *m* polare
p eixo *m* polar
d Polarachse *f*

2087 polar chart
f carte *f* polaire
e carta *f* polar
i carta *f* polare
p carta *f* polar
d Polarkarte *f*

2088 polar climate
f climat *m* polaire
e clima *m* polar
i clima *m* polare
p clima *m* polar
d Polarklima *n*

2089 polar coordinates
f coordonnées *fpl* polaires
e coordenadas *fpl* polares
i coordinate *fpl* polari
p coordenadas *fpl* polares
d Polarkoordinaten *fpl*

2090 polar day
f jour *m* polaire
e día *m* polar
i giorno *m* polare
p dia *m* polar
d Polartag *m*

2091 polar desert
f désert *m* polaire
e desierto *m* polar
i deserto *m* polare
p deserto *m* polar
d Polarwüste *f*

2092 polar distance
f distance *f* polaire
e distancia *f* polar

i distanza *f* polare
p distância *f* polar
d Polardistanz *f*

2093 polar front
f front *m* polaire
e frente *m* polar
i fronte *m* polare
p frente *f* polar
d Polarfront *f*

2094 polar glacier
f glacier *m* polaire
e glaciar *m* polar
i ghiacciaio *m* polare
p glaciar *m* polar
d Polargletscher *m*

2095 polar ice; ice canopy
f glace *f* polaire; plafond *m* de glace
e hielo *m* polar
i ghiaccio *m* polare
p gelo *m* da calota polar
d Polareis *n*; Eisdach *n*

2096 polarimeter
f polarimètre *m*
e polarímetro *m*
i polarimetro *m*
p polarímetro *m*
d Polarimeter *n*

2097 polariscope
f polariscope *m*
e polariscopio *m*
i polariscopio *m*
p polariscópio *m*
d Polariskop *n*

2098 polar low
f dépression *f* polaire
e depresión *f* polar
i depressione *f* polare
p baixa *f* polar
d Polartief *n*

2099 polar night
f nuit *f* polaire
e noche *f* polar
i notte *f* polare
p noite *f* polar
d Polarnacht *f*

2100 polar ray
f rayon *m* polaire
e rayo *m* polar
i raggio *m* polare
p raio *m* polar
d Polstrahl *m*

2101 polar region
f région *f* polaire
e región *f* polar
i regione *f* polare
p região *f* polar
d Polarregion *f*

2102 polar sea
f mer *f* polaire
e mar *m* polar
i mare *m* polare
p mar *m* polar
d Polarsee *f*

2103 polar zone
f zone *f* polaire
e zona *f* polar
i zona *f* polare
p zona *f* polar
d polare Zone *f*

2104 pole height
f hauteur *f* polaire
e altura *f* polar
i altezza *f* polare
p altura *f* polar
d Polhöhe *f*

2105 pollination
f pollinisation *f*
e polinización *f*
i impollinazione *f*
p polinização *f*
d Bestäubung *f*

* **polluted air → 1115**

2106 pollution of ice
f pollution *f* de la glace
e contaminación *f* del hielo
i inquinamento *m* del ghiaccio
p poluição *f* do gelo
d Verschmutzung *f* des Eises

2107 polymeter
f polymètre *m*
e polímetro *m*
i polimetro *m*
p polímetro *m*
d Polymeter *n*

2108 polynya
f polynia *f*
e polínia *f*
i polinia *f*; apertura *f* nel ghiaccio
p polínia *f*
d Polynja *f*

2109 pool
f étang *m*
e laguna *f*
i stagno *m*
p charco *m*
d Teich *m*

* **poor visibility → 1685**

2110 porosity of ice
f porosité *f* de la glace
e porosidad *f* del hielo
i porosità *f* del ghiaccio
p porosidade *f* do gelo
d Eisporosität *f*

2111 porosity of snow
f porosité *f* de la neige
e porosidad *f* de la nieve
i porosità *f* della neve
p porosidade *f* da neve
d Schneeporosität *f*

2112 portable anemometer
f anémomètre *m* portatif
e anemómetro *m* portátil
i anemometro *m* portatile
p anemômetro *m* portátil
d tragbarer Windmesser *m*

2113 portable barometer
f baromètre *m* portatif
e barómetro *m* portátil
i barometro *m* portatile
p barômetro *m* portátil
d Reisebarometer *n*

2114 post-blossom spray
f pulvérisation *f* postflorale
e pulverización *f* postfloral
i trattamento *m* postfloreale
p pulverização *f* pós-floral
d Nachblütespritzung *f*

2115 potential gradient
f gradient *m* de potentiel
e gradiente *m* de potencial
i gradiente *m* di potenziale
p gradiente *m* de potencial
d Potentialgradient *m*

2116 potential temperature
f température *f* potentielle
e temperatura *f* potencial
i temperatura *f* potenziale
p temperatura *f* potencial
d potentielle Temperatur *f*

2117 powder snow
f neige *f* poudreuse
e polvo *m* de nieve
i polvere *f* di neve
p poeira *f* de neve
d Pulverschnee *m*

2118 powdery avalanche
f avalanche *f* poudreuse
e alud *m* de polvo
i valanga *f* di polvere
p avalancha *f* de poeira
d Staublawine *f*

2119 prairie
f prairie *f*
e pradera *f*; llanura *f*
i prateria *f*
p pradaria *f*
d Prärie *f*

2120 precession of the equinoxes
f précession *f* des équinoxes
e precesión *f* de los equinoccios
i precessione *f* di equinozi
p precessão *f* do equinócio
d Präzession *f* der Äquinoktien

* **precipitation → 258**

2121 precipitation area
f zone *f* de précipitation
e zona *f* de precipitación
i zona *f* di precipitazione
p zona *f* de precipitação
d Niederschlagsgebiet *n*

2122 precipitation deficit
f déficit *m* de précipitation
e déficit *m* de precipitación
i deficit *m* di piovosità
p déficit *m* de precipitação
d Niederschlagsdefizit *n*

2123 precipitation surplus
f excès *m* de précipitation
e exceso *m* de precipitación
i eccesso *m* di precipitazioni
p excesso *m* de precipitação
d Niederschlagsüberschuß *m*

2124 precipitation tank
f cuve *f* de précipitation
e tanque *m* de precipitación
i bacino *m* di precipitazione
p tanque *m* de precipitação
d Präzipitationstank *m*

* **prediction → 1100**

2125 preharvest drop
f chute *f* des fruits avant la récolte
e caída *f* ante cosecha
i cascola *f* pre-raccolta
p queda *f* pré-colheita
d vorzeitiger Fruchtfall *m*

2126 pressure
f pression *f*
e presión *f*
i pressione *f*
p pressão *f*
d Druck *m*

* **pressure altimeter → 337**

2127 pressure altitude
f altitude *f* de pression
e altura *f* de presión
i quota *f* di pressione
p altitude *f* de pressão
d Druckhöhe *f*

2128 pressure contour
f isohypse *f*
e contorno *m* de presión
i linea *f* a pressione costante
p linha *f* de pressão constante
d Luftdruckhöhenlinie *f*

* **pressure gradient → 338**

2129 pressure head
f pression *f* hydraulique
e carga *f* hidráulica
i carico *m* idraulico
p altura *f* de pressão
d Wasserdruck *m*

2130 pressure of floating ice
f poussée *f* de la glace flottante
e presión *f* del hielo flotante
i pressione *f* del ghiaccio galleggiante
p pressão *f* do gelo flutuante
d Treibeisdruck *m*

2131 pressure of water vapour; water-vapour pressure
f pression *f* de la vapeur d'eau
e tensión *f* de vapor de agua
i pressione *f* del vapore acqueo
p tensão *f* do vapor d'água
d Wasserdampfdruck *m*

2132 pressure on the environment
f dégradation *f* de l'environnement
e sobrecarga *f* del medio ambiente
i perturbazione *f* dell'ambiente
p degradação *f* do meio ambiente
d Umweltbelastung *f*

2133 pressure pattern flying; aerologation
f vol *m* barique
e navegación *f* isobárica
i aeronavigazione *f* isobarica
p navegação *f* bárica; vôo *m* bárico
d meteorologische Navigation *f*

2134 pressure surge
f augmentation *f* rapide de la pression atmosphérique
e aumento *m* brusco de la presión
i aumento *m* improvviso della pressione
p aumento *m* brusco da pressão atmosférica
d plötzliches Ansteigen *n* des Luftdruckes

2135 pressure tube anemometer
f anémomètre *m* hydrostatique
e anemómetro *m* de tubo de presión
i anemometro *m* a pressione
p anemômetro *m* manométrico
d Staurohranemometer *n*

2136 pressure wave
f onde *f* de pression
e onda *f* de presión
i onda *f* di pressione
p onda *f* de pressão
d Druckwelle *f*

2137 pressurizing
f manutention *f* sous pression
e presionización *f*
i pressurizzazione *f*
p pressurização *f*
d Unterdrucksetzen *n*

2138 prevailing wind; prevailing wind direction
f vent *m* dominant
e viento *m* dominante
i direzione *f* prevalente del vento
p vento *m* dominante
d vorherrschender Wind *m*

* **prevailing wind direction → 2138**

2139 primary air
f air *m* primaire
e aire *m* primario
i aria *f* primaria
p ar *m* primário
d Primärluft *f*

2140 primary triangulation
f triangulation *f* fondamentale
e triangulación *f* principal
i triangolazione *f* di primo ordine
p triangulação *f* principal
d Haupttriangulation *f*

* **primeval forest → 2907**

2141 prismatic astrolabe
f astrolabe *m* à prisme
e astrolabio *m* de prisma
i astrolabio *m* a prisma
p astrolábio *m* de prisma
d Prismenastrolab *m*

2142 private farming
f agriculture *f* privée
e agricultura *f* privada
i agricoltura *f* privata
p agricultura *f* privada
d private Landwirtschaft *f*

2143 probability
f probabilité *f*
e probabilidad *f*
i probabilità *f*
p probabilidade *f*
d Wahrscheinlichkeit *f*

2144 probable error
f erreur *f* probable
e error *m* probable
i errore *m* probabile
p erro *m* provável
d wahrscheinlicher Fehler *m*

2145 prognostic chart
f carte *f* prévue
e carta *f* pronosticadora
i carta *f* prevista
p carta *f* de previsão
d Vorhersagekarte *f*

2146 prominence
f proéminence *f*
e protuberancia *f*
i protuberanza *f*
p protuberância *f*
d Protuberanz *f*

2147 propagation house
f serre *f* de multiplication
e estufa *f* de multiplicación
i serra *f* di moltiplicazione
p estufa *f* de multiplicação
d Vermehrungshaus *n*

2148 propagation of radio waves
f propagation *f* des ondes électromagnétiques
e propagación *f* de ondas de radio
i propagazione *f* delle onde di radio
p propagação *f* de ondas de rádio
d Funkwellenausbreitung *f*

2149 propagation of sound
f propagation *f* du son
e propagación *f* del sonido
i propagazione *f* del suono
p propagação *f* do som
d Schallausbreitung *f*

2150 propagation of sound in the free atmosphere
f propagation *f* du son dans l'air libre
e propagación *f* del sonido en atmósfera libre
i propagazione *f* del suono nell'atmosfera libera
p propagação *f* do som na atmosfera livre
d Schallfortpflanzung *f* in der freien Atmosphäre

2151 propagation of waves
f propagation *f* d'ondes
e propagación *f* de olas
i propagazione *f* di onde
p propagação *f* de ondas
d Wellenfortpflanzung *f*

2152 propagation period
f période *f* de multiplication
e período *m* de propagación
i periodo *m* di propagazione
p período *m* de propagação
d Vermehrungsperiode *f*

2153 protection against erosion
f protection *f* contre l'érosion
e protección *f* contra la erosión
i protezione *f* dall'erosione
p proteção *f* contra a erosão
d Erosionsschutz *f*

*** protection against frost → 1155**

2154 protection forest
f forêt *f* de protection
e bosque *f* de protección
i bosco *m* di protezione
p floresta *f* de proteção
d Bannwald *m*; Schutzwaldung *f*

*** protection of nature → 1901**

2155 protection of produce
f protection *f* des produits
e protección *f* de los productos
i protezione *f* del prodotto
p proteção *f* dos produtos
d Produktschutz *m*

2156 pseudofront
f pseudofront *m*
e seudofrente *m*
i pseudofronte *m*

p pseudo-frente *f*
d Pseudofront *f*

2157 psychrometer
f psychromètre *m*
e psicrómetro *m*
i psicometro *m*
p psicrômetro *m*
d Psychrometer *n*

2158 pulsating arc
f arc *m* pulsant
e arco *m* pulsante
i arco *m* pulsante
p arco *m* pulsante
d pulsierender Bogen *m*

2159 purification
f épuration *f*
e depuración *f*
i depurazione *f*
p purificação *f*
d Reinigung *f*

2160 purple light
f lueur *f* pourprée
e luz *f* purpúrea
i luce *f* purpurea
p luz *f* purpúrea
d Purpurlicht *n*

2161 pyramid iceberg
f iceberg *m* pyramidal
e iceberg *m* piramidal
i iceberg *m* piramidale
p icebergue *m* piramidal
d Pyramidaleisberg *m*

2162 pyrgeometer
f pyrgéomètre *m*
e pirgeómetro *m*
i pirgeometro *m*
p pirgeômetro *m*
d Pyrgeometer *n*

2163 pyrheliometer
f pyrhéliomètre *m*
e pirheliómetro *m*
i pireliometro *m*
p pirheliômetro *m*
d Pyrheliometer *n*

Q

2164 quadrant
f quadrant *m*
e cuadrante *m*
i quadrante *m*
p quadrante *m*
d Quadrant *m*

2165 qualitative cultivation
f culture *f* de qualité
e cultivo *m* cualitativo
i coltura *f* qualitativa
p cultura *f* qualitativa
d Qualitätsanbau *m*

2166 quantity of radiation
f quantité *f* d'énergie rayonnante
e cantidad *f* de radiación
i quantità *f* di radiazione
p quantidade *f* de radiação
d Strahlungsmenge *f*

* **quarter of moon → 1872**

2167 quarter wind
f vent *m* de la hanche
e viento *m* largo
i vento *m* largo
p vento *m* de lado
d Backstagswind *m*

2168 quick freezing
f congélation *f* rapide
e congelación *f* rápida
i congelazione *f* rapida
p congelação *f* rápida
d Schnellgefrieren *n*

R

2169 radar
f radar *m*
e radar *m*
i radar *m*
p radar *m*
d Radar *m*

2170 radar balloon
f radar *m* de ballon-sonde
e globo *m* sonda con radar
i pallone *m* sonda con radar
p balão-sonda *m* de radar
d Pilotballon *m* mit Radar

2171 radar bearing
f relèvement *m* par radar
e marcación *f* por radar
i rilevamento *m* per radar
p marcação *f* por radar
d Radarpeilung *f*

2172 radar system
f système *m* radar
e sistema *m* radar
i sistema *m* radar
p sistema *m* radar
d Radarsystem *n*

2173 radiant *adj*
f radiant
e radiante
i radiante
p radiante
d strahlend

2174 radiant energy
f énergie *f* radiante; énergie *f* rayonnante
e energía *f* radiante
i energia *f* raggiante
p energia *f* radiante
d Strahlungsenergie *f*

2175 radiant flux
f flux *m* énergétique
e flujo *m* energético
i flusso *m* raggiante
p fluxo *m* radiante
d Strahlungsfluß *m*

* **radiant flux density → 2179**

2176 radiant point
f point *m* rayonnant
e punto *m* radiante
i punto *m* radiante
p ponto *m* irradiante
d Strahlungspunkt *m*

2177 radiation
f radiation *f*
e radiación *f*
i radiazione *f*
p radiação *f*
d Strahlung *f*

* **radiation belt → 2180**

2178 radiation fog
f brouillard *m* de rayonnement
e niebla *f* de radiación
i nebbia *f* di radiazione
p nevoeiro *m* formado por irradiação
d Strahlungsnebel *m*

2179 radiation intensity; radiant flux density
f intensité *f* de rayonnement
e intensidad *f* de radiación
i intensità *f* di radiazione
p intensidade *f* de radiação
d Strahlungsintensität *f*

2180 radiation zone; radiation belt
f ceinture *f* de radiation; zone *f* de radiation
e zona *f* de radiación
i cintura *f* di radiazioni
p zona *f* de radiação
d Strahlungsgürtel *m*

2181 radioactive aerosol
f aérosol *m* radioactif
e aerosol *m* radiactivo
i aerosol *m* radioattivo
p aerosol *m* radioativo
d radioaktives Aerosol *n*

2182 radioactive decay
f désagrégation *f* radioactive
e decaimiento *m* radiactivo
i disintegrazione *f* radioattiva
p desintegração *f* radioativa
d radioaktiver Zerfall *m*

2183 radioactive fall-out
f retombée *f* radioactive
e precipitación *f* radiactiva
i precipitazione *f* radioattiva
p precipitação *f* radioativa
d radioaktiver Niederschlag *m*

2184 radioactive snow gauge
f densimètre *m* par mesure de radioactivité de la neige
e nivómetro *m* radiactivo de la nieve
i nivometro *m* radioattivo della neve
p nivômetro *m* radioativo da neve
d Schneedichtemesser *n* von Radioaktivitätsprinzip

2185 radioactive water
f eau *f* radioactive
e agua *f* radiactiva
i acqua *f* radioattiva
p água *f* radioativa
d radioaktives Wasser *n*

2186 radioactivity
f radioactivité *f*
e radiactividad *f*
i radioattività *f*
p radioatividade *f*
d Radioaktivität *f*

2187 radio balloon; radio sonde; radio wind flight
f radiosonde *f*
e globo *m* radiosonda
i pallone *m* radio; radiosonda *m*
p balão *m* com radiossonda; radiosonda *m*
d Radiosonde *f*

* **radiodirection finder → 2192**

2188 radioecology
f radioécologie *f*
e radioecología *f*
i radioecologia *f*
p radioecologia *f*
d Radioökologie *f*

2189 radioelectric storm detection
f détection *f* radioélectrique des tempêtes
e detección *f* radioeléctrica de tempestades
i localizzazione *f* di una perturbazione radioelettrica
p detecção *f* radioelétrica de tempestades
d elektrische Gewitterpeilung *f*

2190 radio fog signals
f signaux *mpl* de brume radioélectriques
e señales *fpl* radioeléctricas de niebla
i segnali *mpl* radiotelegrafici della nebbia
p sinais *mpl* radiotelegráficos em tempo de nevoeiro
d Funknebelsignale *npl*

2191 radiogenic heat
f chaleur *f* radiogénique
e calor *m* radiogénico
i calore *m* radiogenico
p calor *m* radiogênico
d radiogene Wärme *f*

2192 radiogoniometer; radiodirection finder
f radiogoniomètre *m*
e radiogoniómetro *m*
i radiogoniometro *m*
p radiogoniômetro *m*
d Radiogoniometer *n*; Funkpeiler *m*

2193 radiogoniometry
f radiogoniométrie *f*
e radiogoniometría *f*
i radiogoniometria *f*
p radiogoniometria *f*
d Radiogoniometrie *f*

2194 radiometeorograph
f radiométéorographe *m*
e radiometeorógrafo *m*
i radiometeorografo *m*
p radiometeorógrafo *m*
d radiometeorograph *m*

* **radio-receiver aerial → 2228**

* **radio sonde → 2187**

* **radio wind flight → 2187**

2195 radon
f radon *m*
e radón *m*
i radon *m*
p rádon *m*
d Radon *n*

2196 rain
f pluie *f*
e lluvia *f*
i pioggia *f*
p chuva *f*
d Regen *m*

2197 rain band
f bande *f* de la pluie
e banda *f* de absorción de la lluvia
i banda *f* della pioggia
p risca *f* de chuva
d Regenbande *f*

2198 rainbow
f arc-en-ciel *m*
e arco *m* iris
i arcobaleno *m*
p arco-íris *m*
d Regenbogen *m*

2199 rain cloud
f nuage *m* de pluie
e nube *f* de lluvia
i nube *f* di pioggia
p nuvem *f* de chuva
d Regenwolke *f*

2200 raincoat; tight
f imperméable *m*
e impermeable *m*
i impermeabile *m*
p impermeável *m*
d Regenmantel *m*

2201 rain day
f jour *m* de pluie
e día *m* de lluvia
i giorno *m* di pioggia
p dia *m* de chuva
d Regentag *m*

2202 raindrop
f goutte *f* de pluie
e gota *f* de lluvia
i goccia *f* di pioggia
p gota *f* de chuva
d Regentropfen *m*

2203 raindrop impact
f choc *m* de goutte de pluie
e choque *m* de gotas de lluvia
i urto *m* di gocciole di pioggia
p impacto *m* de gotas de chuva
d Stoß *m* der Regentropfen

2204 rainfall
f précipitation *f*
e precipitación *f*
i precipitazione *f*
p precipitação *f*
d Niederschlag *f*

2205 rainfall coefficient
f coefficient *m* pluvial
e coeficiente *m* pluvial
i coefficiente *m* pluviale
p coeficiente *m* pluvial
d Niederschlagsbeiwert *m*

* **rainfall height → 1347**

* **rainfall recorder → 2081**

2206 rain ice; glazed frost
f verglas *m*
e lluvia *f* helada
i ghiaccio *m* di pioggia
p chuva *f* gelada
d Glatteis *n*

* **raining period → 2207**

2207 raining season; raining period
f saison *f* des pluies
e estación *f* de lluvias
i stagione *f* delle piogge
p estação *f* de chuvas
d Regenzeit *f*

2208 rain intensity
f intensité *f* de la pluie
e intensidad *f* de lluvia
i intensità *f* di pioggia
p intensidade *f* de chuva
d Regenstärke *f*

2209 rain proof
f résistance *f* à la pluie
e resistencia *f* a la lluvia
i resistenza *f* alla pioggia
p resistência *f* à chuva
d Regenbeständigkeit *f*

2210 rain shadow
f région *f* abritée de la pluie
e sequía *f* orográfica
i area *f* di poca pioggia
p área *f* protegida das chuvas
d Regenschatten *m*

* **rain shower → 1342**

2211 rainsquall; squall with rain; torrential rain; thick squall
f grain *m* mouillé; pluie *f* torrentielle; grain *m* lourd
e chubasco *m* de lluvia; lluvia *f* torrencial; chubasco *m* pesado
i scroscio *m* di pioggia; piovasco *m*
p tormenta *f* de chuva; chuva *f* torrencial; aguaceiro *m*
d Regenbö *f*; Wolkenbruch *m*; geladene Bö *f*

2212 rainstorm
f tempête *f* de pluie
e borrasca *f* de lluvia
i tempesta *f* di pioggia
p tempestade *f* de chuva
d Regensturm *m*

2213 rainwater
f eau *f* de pluie
e agua *f* pluvial
i acqua *f* piovana
p água *f* de chuva
d Regenwasser *n*

2214 rainy *adj*
f pluvieux
e lluvioso

i piovoso
p chuvoso
d regnerisch

2215 rainy region
f région *f* des pluies
e región *f* de las lluvias
i regione *f* delle piogge
p região *f* das chuvas
d Regenregion *f*

2216 rainy spell
f série *f* de pluie
e período *m* largo de lluvia
i periodo *m* lungo di pioggia
p período *m* longo de chuva
d Regenperiode *f*

2217 rainy weather
f temps *m* pluvieux
e tiempo *m* lluvioso
i tempo *m* piovoso
p tempo *m* chuvoso
d regnerisches Wetter *n*

2218 rainy year
f année *f* pluvieuse
e año *m* lluvioso
i annata *f* piovosa
p ano *m* chuvoso
d regnerisches Jahr *n*

2219 rarefied air
f air *m* raréfié
e aire *m* enrarecido
i aria *f* rarefatta
p ar *m* rarefeito
d verdünnte Luft *f*

2220 rate of icing
f vitesse *f* de givrage
e velocidad *f* de engelamiento
i velocità *f* di formazione del ghiaccio
p velocidade *f* de formação de gelo
d Vereisungsgeschwindigkeit *f*

2221 ravine
(small narrow valley with steeply sloping sides)
f ravin *m*
e barranco *m*
i ravina *f*; gola *f*
p ravina *f*
d Schlucht *f*

2222 raw
f temps *m* froid et humide
e tiempo *m* frío y húmedo
i tempo *m* freddo e umido
p tempo *m* frio e úmido
d rauhes Wetter *n*

2223 ray
f rai *m*
e rayo *m*
i raggio *m*
p raio *m*
d Strahl *m*

2224 Rayleigh number
f nombre *m* de Rayleigh
e número *m* de Rayleigh
i numero *m* di Rayleigh
p número *m* de Rayleigh
d Rayleighzahl *f*

2225 reafforestation
f reforestation *f*
e reforestación *f*
i riforestazione *f*
p reflorestamento *m*
d Wiederaufforstung *f*

* **real height → 2830**

2226 Reaumur temperature scale
f échelle *f* de température Réaumur
e escala *f* de temperatura Reaumur
i scala *f* termometrica Réaumur
p escala *f* de temperatura Réaumur
d Reaumurskala *f*

2227 receiver
f récepteur *m*
e recipiente *m*
i ricevitore *m*
p recepiente *m*
d Empfänger *m*

2228 receiving aerial; radio-receiver aerial
f antenne *f* réceptrice
e antena *f* receptora
i antenna *f* ricevente
p antena *f* receptora
d Empfangsantenne *f*

2229 reclamation of land
f mise *f* en valeur de terres incultes
e puesta *f* en cultivo de nuevas tierras
i bonifica *f* di terreni incolti
p aproveitamento *m* de novas terras
d Gewinnung *f* von Neuland

* **reconstructed glacier → 2243**

2230 recording anemometer
f anémomètre *m* enregistreur
e anemómetro *m* registrador

i anemometro *m* registratore
p anemômetro *m* registrador
d Windschreiber *m*

* **recording barometer** → **327**

2231 recording instrument
f instrument *m* enregistreur
e aparato *m* de registro
i strumento *m* registratore
p instrumento *m* registrador
d Schreiber *m*

* **recording rain gauge** → **2081**

* **recording thermometer** → **2729**

2232 recurvature of storm
f courbure *f* de la trajectoire d'un cyclone
e incurvación *f* de la tempestad
i deviazione *f* della tempesta
p encurvamento *m* da tempestade
d Umbiegung *f* der Sturmbahn

2233 red line
f raie *f* rouge
e línea *f* roja
i riga *f* rossa
p linha *f* vermelha
d rote Linie *f*

2234 reed foghorn
f trompette *f* de brume
e trombeta *f* de niebla
i corno *m* di nebbia
p trompa *f* de palheta
d Nebelhorn *n*

2235 reference ellipsoid
f ellipsoïde *m* de référence
e elipsoide *m* de referencia
i ellissoide *m* di riferimento
p elipsóide *m* de referência
d Bezugsellipsoid *n*

2236 reflection
f réflexion *f*
e reflexión *f*
i riflessione *f*
p reflexão *f*
d Reflexion *f*

2237 reflectometer
f réflectomètre *m*
e reflectómetro *m*
i riflettometro *m*
p refletômetro *m*
d Reflektometer *n*

2238 refract *v*
f réfracter
e refractar
i rifrangere
p refratar
d brechen

2239 refraction
f réfraction *f*
e refracción *f*
i rifrazione *f*
p refração *f*
d Refraktion *f*

2240 refractive *adj*
f réfringent
e refractivo
i rifrangente
p refrativo
d brechend

2241 refractivity
f réfringence *f*
e refractividad *f*
i rifrattività *f*
p refratividade *f*
d Brechungsvermögen *n*

* **refrigerant** → **643**

* **refrigerate** *v* → **642**

2242 refrigeration
f réfrigération *f*
e refrigeración *f*
i refrigerazione *f*
p refrigeração *f*
d Kühlung *f*

2243 regenerated glacier; reconstructed glacier
f glacier *m* régénéré
e glaciar *m* regenerado
i ghiaccio *m* rigenerato
p glaciar *m* regenerado
d regenerierter Gletscher *m*

2244 region
f région *f*
e región *f*
i regione *f*
p região *f*
d Region *f*

2245 region of calms; dead air region
f région *f* des calmes
e región *f* de las calmas
i regione *f* delle calme
p região *f* de calmarias
d Windstillengebiet *n*

2246 region of cyclones
f région *f* de cyclones
e región *f* de ciclones
i regione *f* di cicloni
p região *f* de ciclones
d Zyklonenregion *f*

2247 region of disturbance
f région *f* de perturbation
e zona *f* de perturbación
i zona *f* di perturbazione
p zona *f* de perturbação
d Störungsgebiet *n*

2248 region of trade winds
f région *f* des vents alizés
e región *f* de los vientos alisios
i regione *f* dei venti alisei
p região *f* de ventos alísios
d Passatregion *f*

2249 regression
f régression *f*
e regresión *f*
i regressione *f*
p regressão *f*
d Zurückgehen *n*

2250 regur
(black cotton soil)
f regur *m*
e regur *m*
i regur *m*
p regur *m*
d Regur *m*

2251 relative air humidity
f humidité *f* atmosphérique relative
e humedad *f* relativa del aire
i umidità *f* relativa dell'aria
p umidade *f* relativa do ar
d relative Luftfeuchtigkeit *f*

2252 relative velocity of the wind
f vitesse *f* relative du vent
e velocidad *f* relativa del viento
i velocità *f* relativa del vento
p velocidade *f* relativa do vento
d relative Windgeschwindigkeit *f*

2253 remote sensing
f télédétection *f*
e detección *f* remota
i misurazione *f* a distanza
p detecção *f* remota
d Spüren *n* auf große Entfernung

2254 rendzina
(intrazonal soil formed in humid to semiarid climates)
f rendzine *f*
e rendzina *f*
i rendzina *f*
p rendzina *f*
d Rendzina *f*

2255 renewable raw materials
f matières *fpl* premières renouvelables
e recursos *mpl* naturales renovables
i materie *fpl* prime rinnovabili
p recursos *mpl* naturais renováveis
d nachwachsende Rohstoffe *mpl*

2256 reservoir
f réservoir *m*
e depósito *m*
i serbatoio *m*; magazzino *m*
p reservatório *m*
d Behälter *m*

2257 residual air
f air *m* résiduel
e aire *m* residual
i aria *f* residua
p ar *m* residual
d Residualluft *f*

2258 resistance
f résistance *f*
e resistencia *f*
i resistenza *f*
p resistência *f*
d Widerstandskraft *f*

2259 resistance thermometer
f thermomètre *m* à résistance
e termómetro *m* de resistencia
i termometro *m* a resistenza
p termômetro *m* de resistência
d Widerstandsthermometer *n*

2260 respiration
f respiration *f*
e respiración *f*
i respirazione *f*
p respiração *f*
d Atmung *f*

2261 respiration rate
f intensité *f* respiratoire
e intensidad *f* respiratoria
i intensità *f* di respirazione
p intensidade *f* respiratória
d Atmungsintensität *f*

2262 respiratory heat
f chaleur *f* de respiration
e calor *m* de respiración
i calore *m* di respirazione
p calor *m* de respiração
d Atmungswärme *f*

2263 respiratory insufficiency
f insuffisance *f* respiratoire
e insuficiencia *f* respiratoria
i insufficienza *f* respiratoria
p insuficiência *f* respiratória
d Atmungsinsuffizienz *f*

2264 revolution of the moon
f révolution *f* lunaire
e revolución *f* de la luna
i rivoluzione *f* della luna
p revolução *f* lunar
d Umlauf *m* des Mondes

*** revolving storm → 2808**

2265 rewetting
f réhumidification *f*
e rehumedecimiento *m*
i riumidificazione *f*
p re-umidificação *f*
d Wiederbefeuchtung *f*

2266 Reynolds number
f nombre *m* de Reynolds
e número *m* de Reynolds
i numero *m* di Reynolds
p número *m* de Reynolds
d Reynoldsche Zahl *f*

2267 rhizosphere
f rhizosphère *f*
e rizosfera *f*
i rizosfera *f*
p rizosfera *f*
d Rhizosphäre *f*

*** Rhodesian fever → 894**

2268 rhumb
f rhumb *m*; aire *f* de vent
e rumbo *m*
i rombo *m* di vento
p loxodroma *m*
d Kompaßstrich *m*

2269 ridge of high pressure
f crête *f* anticyclonique
e cresta *f* de alta presión
i cresta *f* di alta pressione
p crista *f* de alta pressão
d Hochdruckrücken *m*

2270 Rift valley fever
f fièvre *f* de la vallée du Rift
e fiebre *f* del valle de Rift
i febbre *f* della valle del Rift
p febre *f* do vale de Rift
d Rifttalfieber *n*

2271 rigidity
f rigidité *f*
e rigidez *f*
i rigidità *f*
p rigidez *f*
d Steifheit *f*

2272 rigidity of snow; strength of snow
f rigidité *f* de la neige
e rigidez *f* de la nieve
i rigidità *f* della neve
p rigidez *f* da neve
d Schneefestigkeit *f*

2273 rime ice
f givre *m* blanc
e cencellada *f*
i ghiaccio *m* granito
p camada *f* de névoa gelada
d Rauhreifvereisung *f*

2274 riparian
f riveraine *m*
e ribereño *m*
i rivierasco *m*; littoraneo *m*
p ribeirinho *m*
d Uferbewohner *m*

2275 ripeness
f maturité *f*
e madurez *f*
i maturità *f*
p maturação *f*
d Reife *f*

2276 ripening
f maturation *f*
e maduración *f*
i maturazione *f*
p amadurecimento *m*
d Reifung *f*

2277 ripple
f risette *f*
e mar *m* rizado
i increspatura *f*
p mar *m* encrespado
d Kräuselung *f*

*** rise → 1493**

2278 rise in temperature; temperature increase
f hausse *f* de la température
e aumento *m* de temperatura
i aumento *m* della temperatura
p aumento *m* de temperatura
d Temperaturzunahme *f*

2279 rising of the water
f montée *f* des eaux
e creciente *f* de agua
i piena *f* di acqua
p elevação *f* das águas
d Steigen *n* des Wassers

2280 rising tide
f marée *f* montante
e marea *f* creciente
i marea *f* montante; marea *f* crescente
p maré *f* crescente
d Tidenstieg *m*

2281 risk of erosion
f risque *m* d'érosion
e riesgo *m* de erosión
i rischio *m* di erosione
p risco *m* de erosão
d Erosionsgefahr *f*

2282 river
f rivière *f*
e río *m*
i fiume *m*
p rio *m*
d Fluß *m*

2283 river drift; diluvium; deluge
f diluvium *m*; déluge *m*
e acarreo *m* fluvial; diluvión *m*; diluvio *m*
i diluviale *m*; diluvium *m*
p dilúvio *m*
d Diluvium *n*

2284 river ice
f glace *f* de rivière
e hielo *m* de río
i fiume *m* di ghiaccio
p gelo *m* de rio
d Flußeis *n*

* **river valley soil → 106**

* **roaring forties → 420**

2285 rocket sounding
f sondage *m* par fusées
e sondeo *m* por medio de cohetes
i sondaggio *m* a razzo
p sondagem *f* com foguete
d Raketensondierung *f*; Raketenaufstieg *m*

2286 rock glacier
f glacier *m* rocheux
e glaciar *m* de roca
i ghiacciaio *m* roccioso
p glaciar *m* rochoso
d Blockgletscher *m*; Steinstrom *m*

2287 roll cumulus
f cumulus *m* en rouleaux
e cúmulo *m* en rollo
i roll-cumulo *m*
p cúmulo *m* em forma de rolo
d Rollkumulus *m*

2288 root
f racine *f*
e raíz *f*
i radice *f*
p raiz *f*
d Wurzel *f*

2289 rotary tidal stream
f courant *m* de marée rotatif
e corriente *f* de marea rotativa
i corrente *f* di marea rotativa
p corrente *f* de maré rotativa
d Mahlstromgezeitenwirbel *m*

2290 rotation of the earth round the sun
f rotation *f* de la terre autour du soleil
e movimiento *m* de la tierra en torno del sol
i moto *m* della terra attorno al sole
p movimento *m* da terra em torno do sol
d Umlauf *m* der Erde um die Sonne

S

2291 Sahara Desert
f Désert *m* saharien
e Desierto *m* del Sahara
i Deserto *m* del Saara
p Deserto *m* do Saara
d Sahara-Wüste *f*

2292 Saint Elmo's fire
f feu *m* Saint-Elme
e fuego *m* de San Telmo
i fuoco *m* di Sant'Elmo
p fogo *m* de Santelmo
d Sankt-Elmsfeuer *n*

2293 Saint Martin's summer; Indian summer
f été *m* de Saint-Martin
e verano *m* de San Martín
i estate *f* di San Martino
p verão *m* de São Martinho
d Altweibersommer *m*

2294 salinity
f salinité *f*
e salinidad *f*
i salinità *f*
p salinidade *f*
d Salzgehalt *m*

2295 salinity of ice
f salinité *f* de la glace
e salinidad *f* del hielo
i salinità *f* di ghiaccio
p salinidade *f* do gelo
d Salzgehalt *m* im Eis

2296 salinometer
f salinomètre *m*
e salinómetro *m*
i salinometro *m*
p salinômetro *m*
d Salzmesser *m*

2297 salubrious climate
f climat *m* salutaire
e clima *m* saludable
i clima *m* salutare
p clima *m* saudável
d Heilklima *n*

2298 sample
f échantillon *m*
e muestra *f*
i campione *m*
p amostra *f*
d Muster *n*

2299 sand bank
f banc *m* de sable
e banco *m* de arena
i banco *m* di sabbia
p banco *m* de areia
d Sandbank *f*

2300 sand pillar
f trombe *f* de sable
e tolvanera *f*
i tromba *f* di sabbia
p coluna *f* de areia
d Staubsäule *f*

2301 sandstorm; dust devil
f tempête *f* de sable
e tempestad *f* de arena; remolino *m* de polvo
i tempesta *f* di sabbia
p tempestade *f* de areia; turbilhão *m* de poeira
d Sandsturm *m*

2302 sandy desert
f désert *m* de sable
e desierto *m* de arena
i deserto *m* di sabbia
p deserto *m* de areia
d Sandwüste *f*

2303 sandy ice
f glace *f* renforcée par du sable
e hielo *m* arenoso
i ghiaccio *m* sabbioso
p gelo *m* arenoso
d Sandeis *n*

* **sargasso → 1288**

2304 Sargasso Sea
f Mer *f* de Sargasses
e Mar *m* de Sargazos
i Mare *m* di Sargassi
p Mar *m* de Sargaços
d Sargassomeer *n*

2305 sastruga; zastruga
(ridge of hard snow formed by action of wind)
f sastruga *f*
e sastruga *f*
i sastruga *f*

p sastruga *f*
d Zastruga *f*

* **satellite** → **219**

2306 satellite observation
f observation *f* du satellite
e observación *f* del satélite
i osservazione *f* del satellite
p observação *f* do satélite
d Satellitbeobachtung *f*

2307 saturate *v*
f saturer
e saturar
i saturare
p saturar
d sättigen

2308 saturated *adj*
f saturé
e saturado
i saturo
p saturado
d gesättigt

2309 saturated adiabatic lapse rate
f gradient *m* adiabatique saturé
e gradiente *m* adiabático saturado
i gradiente *m* adiabatico saturato
p gradiente *m* adiabático saturado
d feuchtadiabatischer Temperaturgradient *m*

2310 saturated zone
f zone *f* imbibée d'eau
e zona *f* humedecida
i zona *f* di saturazione
p zona *f* de saturação
d Vernässungszone *f*

2311 saturation
f saturation *f*
e saturación *f*
i saturazione *f*
p saturação *f*
d Sättigung *f*

2312 saturation deficit
f déficit *m* de saturation
e déficit *m* de saturación
i deficit *m* di saturazione
p déficit *m* de saturação
d Sättigungsdefizit *n*

2313 saturation point
f point *m* de saturation
e punto *m* de saturación
i punto *m* di saturazione
p ponto *m* de saturação
d Sättigungspunkt *m*

2314 saturation vapour pressure
f pression *f* de vapeur saturante
e tensión *f* saturante del vapor
i tensione *f* di vapore di saturazione
p pressão *f* de vapor saturado
d Sättigungsdampfdruck *m*

2315 savanna
f brousse *f*; savane *f*
e sábana *f*
i savana *f*
p savana *f*
d Buschsteppe *f*

2316 scale
f écaille *f*
e escama *f*
i scaglia *f*
p escama *f*
d Schuppe *f*

2317 scale factor
f échelle *f*
e factor *m* de escala
i fattore *m* di scala
p fator *m* de escala
d Maßstabbeiwert *m*

2318 scale height
f hauteur *f* d'échelle
e altura *f* de escala
i altezza *f* di scala; altezza *f* scalare
p altura *f* de escala
d Skalenhöhe *f*

2319 scale of a chart
f échelle *f* de carte
e escala *f* de carta
i scala *f* di carta
p escala *f* de carta
d Maßstab *m* einer Karte

2320 scale of state of sea
f échelle *f* de l'état de la mer
e escala *f* del estado de mar
i scala *f* dello stato del mare
p escala *f* do estado do mar
d Seegangskala *f*

2321 scant *v*
f refuser
e escasearse
i refiutare
p escantear-se
d schralen

* **scarf cloud** → **2894**

2322 scattered clouds
f nuages *mpl* épars
e nubes *fpl* dispersadas
i nubi *fpl* disperse
p nuvens *fpl* dispersas
d Wolkenfetzen *mpl*

2323 scattered starlight
f lumière *f* stellaire diffuse
e luz *f* estelar difusa
i luce *f* stellare diffusa
p luz *f* estelar difusa
d gestreutes Sternlicht *n*

2324 scintillation
f scintillation *f*
e centelleo *m*
i scintillazione *f*
p cintilação *f*
d Szintillation *f*

2325 Scotch mist
f brume *f* écossaise
e neblina *f* escocesa
i precipitazione *f* da nebbia fine
p neblina *f* da escócia
d nasser Nebel *m*

2326 scud *v* **before the storm**
f fuir devant le temps
e huir antes el tiempo
i fuggire davanti al tempo
p fugir antes da tempestade
d vor einem Sturm lenzen

2327 sea
f mer *f*
e mar *m*
i mare *m*
p mar *m*
d See *f*

2328 sea breeze
f brise *f* de mer
e brisa *f* de mar
i brezza *f* di mare
p brisa *f* de mar
d Seewind *m*

2329 sea disturbance
f agitation *f* de la mer
e estado *m* del mar
i burrasca *f* marina
p perturbação *f* do mar
d Seegang *m*

2330 sea fog; sea haze; frost smoke
f brouillard *m* marin; brouillard *m* du matin
e niebla *f* marítima; niebla *f* de mañana
i nebbia *f* marina; nebbia *f* di mattina
p nevoeiro *m* marítimo; neblina *f* da manhã
d Seenebel *m*

* **sea haze → 2330**

2331 sea ice
f glace *f* marine
e hielo *m* marino
i ghiaccio *m* marino
p gelo *m* marinho
d Meereseis *n*

2332 sea level
f niveau *m* de la mer
e nivel *m* del mar
i livello *m* del mare
p nível *m* do mar
d Meeresniveau *n*

2333 seaquake flood wave
f onde *f* marine séismique
e onda *f* sísmica del mar
i onda *f* sismica marina
p onda *f* sísmica do mar
d Erdbebenflutwelle *f*

2334 seasickness
f mal *m* de mer
e mareo *m*
i malattia *f* di mare
p mareio *m*
d Seekrankheit *f*

2335 season
f saison *f*
e estación *f* del año
i stagione *f*
p estação *f*
d Jahreszeit *f*

2336 seasonal *adj*
f saisonnier
e estacional
i stagionale
p sazonal; estacional
d jahreszeitlich

2337 seasonality
f saisonnalité *f*
e estacionalidad *f*
i stagionalità *f*
p sazonalidade *f*
d Saisonalität *f*

2338 seasonal movement
f mouvement *m* saisonnier
e movimiento *m* estacional
i movimento *m* stagionale

p movimento *m* sazonal
d jahreszeitliche Schwankung *f*

2339 seasonal winds
f vents *mpl* saisonniers
e vientos *mpl* estacionales
i venti *mpl* stagionali
p ventos *mpl* sazonais
d saisonmäßige Winde *mpl*

2340 seawater
f eau *f* de mer
e agua *f* de mar
i acqua *f* marina
p água *f* do mar
d Seewasser *n*

2341 secondary air
f air *m* secondaire
e aire *m* secundario
i aria *f* secondaria
p ar *m* secundário
d Sekundärluft *f*

2342 second blooming
f seconde floraison *f*
e segunda floración *f*
i fioritura *f* secundaria
p segunda floração *f*
d Nachblüte *f*

2343 secular magnetic variations
f variations *fpl* magnétiques séculaires
e variaciones *fpl* magnéticas seculares
i variazioni *fpl* magnetiche secolari
p variações *fpl* magnéticas seculares
d säkulare magnetische Variationen *fpl*

2344 secular trend; secular variation
f variation *f* séculaire
e tendencia *f* secular
i variazione *f* secolare
p variação *f* secular
d Säkularschwankung *f*

*** secular variation → 2344**

2345 sediment
f sédiment *m*
e sedimento *m*
i sedimento *m*
p sedimento *m*
d Sediment *n*; Bodensatz *m*

2346 seed orchard
f jardin *m* à graines
e jardín *m* de semillas
i giardino *m* seminato
p pomar *m* para produção de semente
d Samenspendeanlage *f*

2347 seed year
f année *f* de récolte
e año *m* de producción de semilla
i anno *m* di produzione del seme
p ano *m* de safra
d Samenjahr *n*

*** seism → 877**

2348 seismic focus
f hypocentre *m*
e foco *m* sísmico
i ipocentro *m*
p foco *m* sísmico
d Hypozentrum *n*

2349 seismic wave
f onde *f* séismique
e ola *f* sísmica
i onda *f* sismica
p onda *f* sísmica
d Erdbebenwelle *f*

2350 seismogram
f séismogramme *m*
e sismograma *m*
i sismogramma *m*
p sismograma *m*; registro *m* de sismógrafo
d Seismogramm *n*

2351 seismograph
f séismographe *m*
e sismógrafo *m*
i sismografo *m*
p sismógrafo *m*
d Seismograph *m*

2352 selective absorption
f absorption *f* sélective
e absorción *f* selectiva
i assorbimento *m* selettivo
p absorção *f* seletiva
d selektive Absorption *f*

2353 semi-diurnal tides
f marées *fpl* semi-diurnes
e mareas *fpl* semidiurnas
i maree *fpl* semidiurne
p marés *fpl* semidiurnas
d halbtägige Gezeiten *fpl*

2354 semi-diurnal wave
f onde *f* semi-diurne
e onda *f* semidiurna
i onda *f* semidiurna
p onda *f* semidiurna
d halbtägige Welle *f*

2355 sensor
f capteur *m*
e sensor *m*
i sensore *m*
p sensor *m*
d Sensor *m*

2356 September
f septembre *m*
e septiembre *m*
i settembre *m*
p setembro *m*
d September *m*

2357 serene weather; fine weather
f temps *m* serein; beau temps *m*
e tiempo *m* sereno
i tempo *m* sereno
p tempo *m* sereno
d heiteres Wetter *n*; schönes Wetter *n*

2358 serial temperatures
f températures *fpl* en série
e temperaturas *fpl* en serie
i temperature *fpl* seriali
p temperaturas *fpl* em série
d Serientemperaturen *fpl*

2359 set of tide
f sens *m* de la marée
e sentido *m* de la marea
i direzione *f* della corrente di marea
p sentido *m* da maré
d Stromrichtung *f*

2360 settled weather
f temps *m* fait
e tiempo *m* estable
i tempo *m* stabile
p tempo *m* estável
d beständiges Wetter *n*

2361 shade *v*
f ombrer
e sombrear
i ombreggiare
p sombrear
d beschatten

2362 shade temperature; temperature in the shade
f température *f* sous abri
e temperatura *f* a la sombra
i temperatura *f* all'ombra
p temperatura *f* à sombra
d Schattentemperatur *f*

* **shape of the earth → 1037**

2363 sheed lighting
f éclair *m* diffus
e relámpago *m* difuso
i bagliore *m*
p relâmpago *m* difuso
d Flächenblitz *m*

* **shelter belt → 3034**

2364 shelter *v* **from the wind**
f mettre à l'abri du vent
e poner al abrigo de viento
i mettere al riparo di vento
p pôr ao abrigo do vento
d vom Winde schützen

2365 shield
f bouclier *m*
e escudo *m* protector
i scudo *m*
p escudo *m* protetor
d Schild *n*

2366 shield volcano
f volcan *m* bouclier
e volcán *m* en escudo
i vulcano *m* a scudo
p vulcão *m* em escudo
d Schildvulkan *m*

2367 shifting cultivation
f agriculture *f* itinérante
e cultivo *m* por rozas
i sfruttamento *m* silvo-agricolo migratorio
p agricultura *f* itinerante
d wanderende Rodungskultur *f*

2368 shore
f rivage *m*
e ribera *f* de mar
i riva *f* del mare
p litoral *m*
d Meeresufer *n*

2369 shore line
f ligne *f* côtière
e línea *f* de costa
i linea *f* di costa
p linha *f* de costa
d Küstenlinie *f*

2370 short-day treatment
f traitement *m* de jours courts
e tratamiento *m* de día corto
i trattamento *m* a giorno corto
p tratamento *m* de dias curtos
d Kurztagsbehandlung *f*

* **short-period forecast → 2371**

2371 short-range forecast; short-period forecast
f prévision *f* à courte échéance
e pronóstico *m* a corto plazo
i previsione *f* a breve scadenza
p previsão *f* a curto prazo
d kurzfristige Prognose *f*; Kurzfristprognose *f*

2372 short-range weather forecast
f prévision *f* du temps à court terme
e previsión *f* del tiempo a corto alcance
i previsione *f* del tempo a breve termine
p previsão *f* do tempo a curto prazo
d kurzfristige Wettervorhersage *f*

2373 short sea
f mer *f* courte
e marejada *f*
i mare *m* corto
p marejada *f*
d kurze See *f*

2374 short waves; high frequency waves
f ondes *fpl* courtes
e ondas *fpl* cortas
i onde *fpl* corte
p ondas *fpl* curtas
d Kurzwellen *fpl*

2375 shower of volcanic dust
f chute *f* de cendres
e lluvia *f* de ceniza
i pioggia *f* di cenere
p chuva *f* de cinzas
d Aschenregen *m*

2376 shuga
f shuga *f*
e shuga *f*
i ghiaccio *m* spugnoso
p gelo *m* esponjoso
d Eisbrei *m*; Matscheis *n*

2377 Siberian anticyclone
f anticyclone *m* sibérien
e anticiclón *m* siberiano
i anticiclone *m* siberiano
p anticiclone *m* siberiano
d siberischer Antizyklon *m*

2378 sidereal day
f jour *m* sidéral
e día *m* sidéreo
i giorno *m* sidereo
p dia *m* sideral
d Sterntag *m*

2379 sidereal time
f temps *m* sidéral
e tiempo *m* sidéreo
i tempo *m* sidereo
p tempo *m* sideral
d Sternzeit *f*

2380 sidereal year
f année *f* sidérale
e año *m* sidéreo
i anno *m* sidereo
p ano *m* sideral
d Sternjahr *n*

2381 sierozem
(grey desert soil)
f sierozem *m*; sérozem *m*
e sierosem *m*
i sierozem *m*
p sierozém *m*
d Sierozem *m*

2382 silt *v*
f colmater
e colmatarse
i alluvionare
p assorear
d anschlämmen

2383 silting
f colmatage *m*
e colmateo *m*
i colmata *f*
p aluvião *m*
d Auflandung *f*

2384 silver iodide
f iodure *m* d'argent
e yoduro *m* de plata
i ioduro *m* di argento
p iodeto *m* de prata
d Silberjodid *n*

2385 silver thaw
f dégel *m* argenté
e escarcha *f* plateada
i galaverna *f*
p degelo *m* prateado
d Reif *m*

2386 sima
(the lower layer of the eart's crust)
f sima *f*
e sima *f*
i sima *f*
p sima *f*
d Sima *n*

2387 simoon
(warm, dry, dust-laden violent wind blowing from the deserts of Arabia and Africa)
f simoun *m*
e simún *m*

i simun *m*
p simum *m*
d Samum *m*

* **single harvest** → **2388**

2388 single pick; single harvest
f récolte *f* unique
e recolección *f* única
i raccolta *f* unica
p colheita *f* única
d Einmalernte *f*

2389 singularity
f singularité *f*
e singularidad *f*
i singolarità *f*
p singularidade *f*
d Singularität *f*

2390 sinus iridum
f golfe *m* des iris
e bahía *f* de los arcos iris
i golfo *m* delle iridi
p enseada *f* dos arco-íris
d Regenbogenbucht *f*

2391 siphon barometer
f baromètre *m* à siphon
e barómetro *m* de sifón
i barometro *m* a sifone
p barômetro *m* de sifão
d Heberluftdruckmesser *m*

2392 siphon tube
f tube *m* de siphon
e tubo *m* sifón
i tubo *m* a sifone
p tubo *m* de sifão
d Heberrohr *n*

2393 sirocco
(warm dry, south or southeasterly wind blowing from high land of Sahara Desert)
f sirocco *m*
e siroco *m*
i scirocco *m*
p siroco *m*
d Schirokko *m*

2394 size of particles
f grandeur *f* des particules
e tamaño *m* de los granos
i grandezza *f* delle particelle
p tamanho *m* das partículas
d Partikelgröße *f*

2395 skip distance
f portée *f* minimum
e límite *m* exterior de la zona de silencio
i intervallo *m* di salto
p zona *f* de salto
d Sprungentfernung *f*

2396 sky
f ciel *m*
e cielo *m*
i cielo *m*
p céu *m*
d Himmel *m*

2397 sky chart
f carte *f* céleste
e carta *f* del cielo
i carta *f* del cielo
p carta *f* do céu
d Sternkarte *f*

* **sky wave** → **2513**

2398 slightly agitated air
f air *m* modérément en mouvement
e aire *m* poco agitado
i aria *f* leggermente mossa
p ar *m* pouco agitado
d mäßig bewegte Luft *f*

2399 sling thermometer
f thermomètre-fronde *m*
e termómetro *m* honda
i termometro *m* a fionda
p termômetro *m* funda
d Schleuderthermometer *n*

2400 slush
f gadoue *f*
e pasta *f*; grumo *m*
i ghiaccio *m* a frammenti; fango *m* di neve
p pasta *f* de gelo; neve *f* semiderretida
d Matsch *m*; Schneeschlamm *m*

2401 small floe
f floe *m*
e carámbanos *m*
i piccolo *m* ghiaccio
p bloco *m* de gelo flutuante
d Eisscholle *f*

2402 small winter moth; mottled umber moth
f petite phalène *f* hiémale; hibernie *f* défeuillante
e falena *f* invernal *(de los frutales)*
i falena *f* invernale; falena *f* sfogliatrice
p falena *f* invernal
d Frostspanner *m*

2403 smoke
f fumée *f*
e humo *m*

i fumo *m*
p fumo *m*
d Rauch *m*

2404 smoke cloud
f nuage *m* de fumée
e nube *f* de humo
i nube *f* di fumo
p nuvem *f* de fumaça
d Rauchwolke *f*

2405 smoke formation
f formation *f* de fumée
e formación *f* de humo
i formazione *f* di fumo
p formação *f* de fumaça
d Rauchbildung *f*

2406 smoke trail
f traînée *f* de fumée
e reguero *m* de humo
i striscia *f* di fumo
p rastrilho *m* de fumaça
d Rauchspur *f*

2407 snow
f neige *f*
e nieve *f*
i neve *f*
p neve *f*
d Schnee *m*

2408 snow accumulation
f accumulation *f* de la neige
e acumulación *f* de la nieve
i accumulazione *f* della neve
p acumulação *f* de neve
d Schneeakkumulation *f*

2409 snow avalanche
f avalanche *f* de neige
e alud *m* de nieve
i valanga *f* di neve
p avalancha *f* de neve
d Schneelawine *f*

2410 snowball
f viorne *f*; boule *f* de neige
e bola *f* de nieve
i palla *f* di neve
p noveleiro *m*
d Schneeball *m*

*** snow bank → 2417**

2411 snow blindness
f éblouissement *m* nival
e oftalmia *f* de nieve
i cecità *f* da neve
p oftalmia *f* da neve
d Schneeblindheit *f*

2412 snow bridge
f pont *m* de neige
e puente *f* de nieve
i ponte *f* di neve
p ponte *f* de neve
d Schneebrücke *f*

2413 snow cat; snow trac
f chenillette *f* pour pistes de neige
e vehículo *m* de oruga
i gatto *m* delle nevi
p veículo *m* para pistas de neve
d Schneekatze *f*

2414 snow cover; snow lying
f couverture *f* de neige
e cubierto *m* de nieve
i copertura *f* di neve
p cobertura *f* de neve; neve *f* jacente
d Schneedecke *f*; Schneelage *f*

2415 snow crust
f croûte *f* de neige
e crosta *f* de nieve
i crosta *f* di neve
p crosta *f* de neve
d vereiste Schneeoberfläche *f*

2416 snow crystal
f cristal *m* de neige
e cristal *m* de nieve
i cristallo *m* di neve
p cristal *m* de neve
d Schneekristall *n*

2417 snowdrift; snow bank
f monceau *m* de neige; congère *f*
e montón *m* de nieve; banco *m* de nieve
i cumulo *m* di neve; banco *m* di neve
p neve *f* acumulada; banco *m* de neve
d Schneewehe *f*; Schneewalle *m*

2418 snowdrop
f perce-neige *m*
e rompenieve *m*; galanto *m* de nieve
i bucaneve *m*
p campainha *f* branca
d Schneeglöckchen *n*

2419 snow dune
f dune *f* de neige
e duna *f* de nieve
i duna *f* di neve
p duna *f* de neve
d Schneedüne *f*

2420 snowfall
f précipitation *f* nivale
e nevada *f*
i precipitazione *f* nevosa
p precipitação *f* de neve
d Schneefall *m*

2421 snow fence
f barrière *f* contre la neige
e barrera *f* contra la nieve
i steccato *m* antineve
p barreira *f* contra a neve
d Schneeschutzzaun *m*

2422 snowflake; flake of snow; snow grain
f flocon *m* à voile; grésil *m*
e copo *m* de nieve
i fiocco *m* di neve
p floco *m* de neve; grão *m* de neve
d Schneeflocke *f*; Schneekorn *n*

2423 snow gauge
f tube *m* à densité
e nivómetro *m*
i nivometro *m*
p nivômetro *m*
d Schneemesser *m*

2424 snow glacier
f glacier *m* de névé
e glaciar *m* de nieve
i nevaio *m*
p glaciar *m* de neve
d Firngletscher *m*

* **snow grain → 2422**

2425 snow line
f limite *f* des neiges
e límite *m* de las nieves
i limite *m* delle nevi
p linha *f* de neve
d Schneegrenze *f*

2426 snow load
f charge *f* de neige
e carga *f* de nieve
i carico *m* di neve
p carga *f* devida à neve
d Schneelast *f*

* **snow lying → 2414**

2427 snow melt
f fusion *f* de la neige
e fusión *f* de la nieve
i fusione *f* della neve
p fusão *f* da neve
d Schneeschmelze *f*

2428 snow-melting limit
f limite *f* de fusion de la neige
e límite *m* de fusión de la nieve
i limite *m* di fusione della neve
p limite *m* de fusão da neve
d Schneeschmelzgrenze *f*

2429 snow mould
f moisissure *f* de la neige
e moho *m* de la nieve
i muffa *f* delle nevi
p mofo *m* da neve
d Schneeschimmel *m*

2430 snow roller
f rouleau *m* de neige
e rollo *m* de nieve
i cilindro *m* di neve
p rolo *m* de neve
d Schneeroller *m*

2431 snow sky
f reflet *m* des neiges
e cielo *m* de nieve
i cielo *m* da neve
p reflexo *m* da neve *(no céu)*
d Schneehimmel *m*

2432 snow slush
f slush *m*
e hongo *m* de nieve
i fungo *m* di neve
p fungo *m* de neve
d Schneebrei *m*

2433 snow squall; blizzard
f grain *m* accompagné de neige
e chubasco *m* de nieve
i piovasco *m* con neve
p tormenta *f* de neve
d Schneebö *f*; Blizzard *m*

2434 snow storm
f tempête *f* de neige
e tormenta *f* de nieve
i tempesta *f* di neve
p tempestade *f* de neve
d Schneesturm *m*

* **snow trac → 2413**

2435 snow water
f eau *f* provenant des neiges
e agua *f* del deshielo
i acqua *f* di disgelo
p água *f* do degelo
d Schneewasser *n*

2436 soft hail; graupel
f neige *f* roulée; grésil *m*
e granizo *m* blando
i grandine *f* nevosa
p granizo *m* miúdo
d Graupel *m*

2437 soil; earth
f sol *m*; terre *f*
e suelo *m*; tierra *f*
i suolo *m*; terra *f*
p solo *m*; terra *f*
d Boden *m*; Erde *f*

2438 soil aeration
f aération *f* du sol
e aireación *f* del suelo
i aerazione *f* del terreno
p arejamento *m* do solo
d Bodenbelüftung *f*

2439 soil characteristic
f caractéristique *f* du sol
e característica *f* del suelo
i caratteristica *f* del terreno
p característica *f* do solo
d Bodeneigenschaft *m*

2440 soil condition; state of ground
f état *m* du sol
e condición *f* del suelo
i condizione *f* del suolo
p condição *f* do solo
d Bodenzustand *m*

2441 soil conservation
f conservation *f* des sols
e conservación *f* de suelos
i conservazione *f* del terreno
p conservação *f* do solo
d Bodenerhaltung *f*

2442 soil contamination; soil pollution
f contamination *f* du sol; pollution *f* du sol
e contaminación *f* del suelo
i infezione *f* del terreno
p poluição *f* do solo
d Bodenverseuchung *f*

* **soil creep → 661**

2443 soil degeneration
f dégénérescence *f* du sol
e degradación *f* del suelo
i degenerazione *f* del terreno
p degradação *f* do solo
d Bodendegeneration *f*

2444 soil density
f densité *f* du sol
e densidad *f* del suelo
i densità *f* del terreno
p densidade *f* do solo
d Bodendichte *f*

2445 soil dressing; soil fertilization
f fertilisation *f* du sol
e fertilización *f* del suelo
i concimazione *f* del suolo
p fertilização *f* do solo
d Bodendüngung *f*

2446 soil exhaustion
f épuisement *m* du sol
e agotamiento *m* del suelo
i esaurimento *m* del suolo
p esgotamento *m* do solo
d Bodenerschöpfung *f*

2447 soil fertility
f fertilité *f* du sol
e fertilidad *f* de suelo
i fertilità *f* del suolo
p fertilidade *f* do solo
d Bodenfruchtbarkeit *f*

* **soil fertilization → 2445**

* **soil flow → 2487**

2448 soil formation; soil genesis; pedogenesis
(the formation and development of soil from parent material)
f développement *m* du sol; formation *f* du sol
e formación *f* del suelo
i formazione *f* del suolo; genesi *f* del suolo
p formação *f* do solo; pedogênese *f*
d Bodenbildung *f*

* **soil frost → 1276**

2449 soil fumigation
f fumigation *f* du sol
e fumigación *f* del suelo
i fumigazione *f* del terreno
p fumigação *f* do solo
d Bodenbegasung *f*

* **soil genesis → 2448**

2450 soil heat; soil temperature
f chaleur *f* de fond; température *f* du sol
e calor *m* del suelo; temperatura *f* del suelo
i calore *m* del terreno; temperatura *f* del terreno
p calor *m* do solo; temperatura *f* do solo
d Bodenwärme *f*

2451 soil heat flux
f flux *m* de chaleur dans le sol
e flujo *m* de calor en el suelo
i flusso *m* radiante del terreno
p fluxo *m* térmico do solo
d Bodenwärmestrom *m*

2452 soil heating; floor heating
f chauffage *m* du sol
e calefacción *f* del suelo; calentamiento *m* del suelo
i riscaldamento *m* basale; riscaldamento *m* del terreno
p aquecimento *m* do solo
d Bodenheizung *f*

2453 soil improvement
f amélioration *f* du sol
e mejora *f* del suelo
i miglioramento *m* del suolo
p melhoramento *m* do solo
d Bodenverbesserung *f*

* **soil loosening → 973**

2454 soil management
f travail *m* du sol
e tratamiento *m* del suelo
i lavore *m* del suolo
p tratamento *m* do solo
d Bodenbearbeitung *f*

2455 soil moisture
f humidité *f* du sol
e humedad *f* del suelo
i umidità *f* della terra
p umidade *f* do solo
d Bodenfeuchtigkeit *f*

2456 soil moisture probe
f sonde *f* d'humidimètre du sol
e sonda *f* de valorímetro de la humedad del suelo
i sonda *f* d'umidimetro del suolo
p sonda *f* de umidímetro do solo
d Bodenfeuchtigkeitssonde *f*

2457 soil moisture tension
f tension *f* de l'eau du sol
e tensión *f* de humedad del suelo
i tensione *f* idrica del terreno
p tensão *f* da umidade do solo
d Bodenwasserspannung *f*

2458 soil parasite
f parasite *m* du sol
e parásito *m* del suelo
i parassita *m* del terreno
p parasita *m* do solo
d Bodenschädling *m*

* **soil pollution → 2442**

2459 soil porosity
f porosité *f* du sol
e porosidad *f* del suelo
i porosità *f* del terreno
p porosidade *f* do solo
d Bodenporosität *f*

2460 soil preparation
f soins *mpl* du sol
e preparación *f* del suelo
i preparazione *f* del suolo
p preparação *f* do solo
d Bodenpflege *f*

2461 soil salinization
f salinisation *f* du sol
e salinización *f* del suelo
i salinizzazione *f* del terreno; saturazione *f* salina del terreno
p salinização *f* do solo
d Versalzung *f* des Bodens

2462 soil science; pedology
f science *f* du sol; pédologie *f*
e ciencia *f* del suelo; pedología *f*
i scienza *f* del suolo; pedologia *f*
p ciência *f* do solo; pedologia *f*
d Bodenkunde *f*; Pedologie *f*

2463 soil sterilization
f stérilisation *f* du sol
e esterilización *f* del suelo
i sterilizzazione *f* del terreno
p esterilização *f* do solo
d Bodensterilisation *f*

2464 soil surface
f surface *f* du sol
e superficie *f* del suelo
i superficie *f* del terreno
p superfície *f* do solo
d Bodenoberfläche *f*

2465 soil survey
f cartographie *f* pédologique
e cartografía *f* del suelo
i cartografia *f* del suolo
p cartografia *f* pedológica
d Bodenkartierung *f*

* **soil temperature → 2450**

2466 solano
(hot, suffocant summer wind in the southeast coast of Spain)
f solano *m*
e solano *m*
i solano *m*
p solano *m*
d Solano *m*

2467 solar *adj*
f solaire
e solar
i solare
p solar
d Sonnen...

2468 solar activity
f activité *f* solaire
e actividad *f* solar
i attività *f* solare
p atividade *f* solar
d Sonnentätigkeit *f*

2469 solar aureole; solar corona
f couronne *f* solaire
e corona *f* solar
i corona *f* solare
p coroa *f* solar
d Sonnenkranz *m*; Sonnenkorona *f*

2470 solar constant
f constante *f* solaire
e constante *f* solar
i costante *f* solare
p constante *f* solar
d Solarkonstante *f*

* **solar corona → 2469**

2471 solar cross
f croix *f* solaire
e cruz *f* solar
i croce *f* solare
p cruz *f* solar
d Sonnenkreuz *n*

2472 solar day
f jour *m* solaire
e día *m* solar
i giorno *m* solare
p dia *m* solar
d Sonnentag *m*

2473 solar depression
f dépression *f* solaire
e depresión *f* solar
i depressione *f* solare
p depressão *f* solar
d Sonnendepression *f*

2474 solar eclipse
f éclipse *f* de soleil
e eclipse *m* solar
i eclisse *f* solare
p eclipse *m* solar
d Sonnenfinsternis *f*

2475 solar flare
f éruption *f* solaire
e erupción *f* solar
i brillamento *m* solare
p erupção *f* solar; erupção *f* cromosférica
d Sonneneruption *f*

2476 solar flood
f flux *m* solaire
e flujo *m* solar
i flusso *m* solare
p fluxo *m* solar
d Solarflut *f*

2477 solar heating
f chauffage *m* solaire
e calefacción *f* solar
i riscaldamento *m* solare
p aquecimento *m* solar
d Solarheizung *f*

2478 solarimeter
f solarimètre *m*
e solarímetro *m*
i solarimetro *m*
p solarímetro *m*
d Sonnenstrahlungsmesser *m*

2479 solarization; sunstroke
f solarisation *f*
e insolación *f*
i insolazione *f*
p exposição *f* ao sol
d Solarisation *f*

2480 solar prominence
f proéminence *f* solaire
e protuberancia *f* solar
i protuberanza *f* solare
p protuberância *f* solar
d Sonnenprotuberanz *f*

2481 solar radiation; insolation
f rayonnement *m* solaire
e radiación *f* solar
i radiazione *f* solare
p radiação *f* solar
d Sonnenstrahlung *f*

2482 solar sensor
f détecteur *m* solaire
e sensor *m* solar

i sensore *m* solare
p sensor *m* solar
d Sonnensensor *m*

2483 solar spectrum
f spectre *m* solaire
e espectro *m* solar
i spettro *m* solare
p espectro *m* solar
d Sonnenspektrum *n*

2484 solar year
f année *f* solaire
e año *m* solar
i anno *m* solare
p ano *m* solar
d Sonnenjahr *n*

2485 solenoid
f solénoïde *m*
e solenoide *m*
i solenoide *m*
p solenóide *m*
d Solenoid *n*

2486 solid particle
f particule *f* solide
e partícula *f* sólida
i particella *f* solida
p partícula *f* sólida
d feste Partikel *f*; festes Teilchen *n*

2487 solifluction; soil flow
f solifluxion
e solifluxión *f*
i soliflusso *m*
p solifluxão *f*
d Solifluktion *f*

2488 solstice
(the two instants in the year when the sun reaches a maximum north or south inclination)
f solstice *m*
e solsticio *m*
i solstizio *m*
p solstício *m*
d Solstitium *n*; Sonnenwende *f*

2489 solsticial colure
f colure *m* solsticial
e coluro *m* solsticial
i coluro *m* solstiziale
p coluro *m* solsticial
d Solstitialkolur *m*

2490 solsticial point
f point *m* solsticial
e punto *m* solsticial
i punto *m* solstiziale
p ponto *m* solsticial
d Solstitialpunkt *m*

2491 solsticial tide
f marée *f* de solstice
e marea *f* de solsticio
i marea *f* di solstizio
p maré *f* de solstício
d Sonnenwendegezeiten *fpl*

2492 Somali Current
f Courant *m* de la Somalie
e Corriente *f* de Somalia
i Corrente *f* della Somalia
p Corrente *f* da Somalia
d Somalistrom *m*

2493 soot
f suie *f*
e hollín *m*
i fuliggine *f*
p fuligem *f*
d Ruß *m*

2494 sound anemometer
f anémomètre *m* acoustique
e anemómetro *m* acústico
i anemometro *m* acustico
p anemômetro *m* acústico
d Schallwindmesser *m*

2495 sounding
f sondage *m*
e sondeo *m*
i sondaggio *m*
p sondagem *f*
d Sondierung *f*

*** sounding balloon → 316**

2496 sound ray
f rayon *m* sonore
e rayo *m* sonoro
i raggio *m* sonoro
p raio *m* sonoro
d Schallstrahl *m*

2497 sound wave
f onde *f* sonore
e onda *f* sonora
i onda *f* sonora
p onda *f* sonora
d Schallwelle *f*

2498 source of heat
f source *f* de chaleur
e fuente *f* de calor
i sorgente *f* di calore

p fonte *f* de calor
d Wärmequelle *f*

2499 source region
f région *f* d'origine
e área *f* de origen
i regione *f* di sorgenti
p zona *f* de origem
d Ursprungsgegend *f*

2500 south
f sud *m*
e sur *m*
i sud *m*
p sul *m*
d Süden *m*

2501 southeast monsoon
f mousson *f* du sud-est
e monzón *m* del sudeste
i monsone *m* da sud-est
p monção *f* de sudeste
d Südostmonsun *m*

2502 southeast trade wind
f alizé *m* du sud-est
e alisio *m* del sudeste
i aliseo *m* da sud-est
p alísio *m* de sudeste
d Südostpassat *m*

2503 south equatorial current
f courant *m* équatorial sud
e corriente *f* ecuatorial sud
i corrente *f* equatoriale meridionale
p corrente *f* equatorial sul
d Südäquatorialstrom *m*

2504 southern celestial hemisphere
f hémisphère *m* céleste austral
e hemisferio *m* celeste austral
i emisfero *m* celeste australe
p hemisfério *m* celeste austral
d südliche Himmelshalbkugel *f*

2505 southern cross
f croix *f* du sud
e cruz *f* del sud
i croce *f* da sud
p cruz *f* do sul
d Südkreuz *n*

2506 southern hemisphere
f hémisphère *m* austral
e hemisferio *m* austral
i emisfero *m* australe
p hemisfério *m* austral
d südliche Halbkugel *f*

2507 south magnetic latitude
f latitude *f* magnétique sud
e latitud *f* magnética sud
i latitudine *f* magnetica sud
p latitude *f* magnética sul
d südmagnetische Breite *f*

2508 south meridian
f méridien *m* austral
e meridiano *m* sud
i meridiano *m* meridionale
p meridiano *m* sul
d Südmeridian *m*

2509 south pole
f pôle *m* sud
e polo *m* sud
i polo *m* sud
p pólo *m* sul
d Südpol *m*

2510 south westerly wind
f vent *m* du sud-ouest
e ábrego *m*
i libeccio *m*; vento *m* da sud-ovest
p ábrego *m*; vento *m* de sudoeste
d Südwestwind *m*

2511 southwest monsoon
f mousson *f* du sud-ouest
e monzón *m* del sudoeste
i monsone *m* da sud-ovest
p monção *f* de sudoeste
d Südwestmonsun *m*

2512 south wind
f vent *m* du sud
e viento *m* sur
i vento *m* da sud
p vento *m* sul
d Südwind *m*

2513 space wave; sky wave
f onde *f* d'espace
e ola *f* espacial
i onda *f* spaziale
p onda *f* espacial
d Raumwelle *f*

2514 specific atmospheric humidity
f humidité *f* atmosphérique spécifique
e humedad *f* atmosférica específica
i umidità *f* atmosferica specifica
p umidade *f* atmosférica específica
d spezifische Luftfeuchtigkeit *f*

2515 specific gravity
f poids *m* spécifique
e peso *m* específico

i peso *m* specifico
p peso *m* específico
d spezifisches Gewicht *n*

2516 specific heat
f chaleur *f* spécifique; chaleur *f* massique
e calor *m* específico
i calore *m* specifico relativo
p calor *m* específico relativo
d spezifische Wärme *f*

2517 specific humidity
f humidité *f* spécifique
e humedad *f* específica
i umidità *f* specifica
p umidade *f* específica
d spezifische Feuchtigkeit *f*

2518 spectrogram
f spectrogramme *m*
e espectrograma *m*
i spettrogramma *m*
p espectrograma *m*
d Spektrogramm *n*

2519 spectrograph
f spectrographe *m*
e espectrógrafo *m*
i spettrografo *m*
p espectrógrafo *m*
d Spektrograph *m*

2520 spectroheliograph
f spectrohéliographe *m*
e espectroheliógrafo *m*
i spettroeliografo *m*
p espectroheliógrafo *m*
d Spektroheliograph *m*

2521 spectrophotometer
f spectrophotomètre *m*
e espectrofotómetro *m*
i spettrofotometro *m*
p espectrofotômetro *m*
d Spektralphotometer *n*

2522 spectroscope
f spectroscope *m*
e espectroscopio *m*
i spettroscopio *m*
p espectroscópio *m*
d Spektroskop *n*

2523 spectroscopic *adj*
f spectroscopique
e espectroscópico
i spettroscopico
p espectroscópico
d spektroskopisch

2524 spectroscopy
f spectroscopie *f*
e espectroscopia *f*
i spettroscopia *f*
p espectroscopia *f*
d Spektroskopie *f*

2525 spectrum
f spectre *m*
e espectro *m*
i spettro *m*
p espectro *m*
d Spektrum *n*

*** spirit thermometer → 101**

2526 sporadic avalanche
f avalanche *f* sporadique
e alud *m* esporádico
i valanga *f* sporadica
p avalancha *f* esporádica
d sporadische Lawine *f*

2527 sporadic E layer
(layer of ionization within the E region)
f couche *f* E sporadique
e capa *f* E esporádica
i strato *m* E sporadico
p camada *f* E esporádica
d sporadische E-Schicht *f*

2528 spraying
f pulvérisation *f*
e pulverización *f*
i irrorazione *f*
p pulverização *f*
d Spritzen *n*

2529 spray irrigation
f irrigation *f* par aspersion
e riego *m* por aspersión; riego *m* por lluvia artificial
i irrigazione *f* a pioggia
p irrigação *f* através de chuva artificial
d künstliche Beregnung *f*

*** spray region → 997**

2530 spring
f printemps *m*
e primavera *f*
i primavera *f*
p primavera *f*
d Frühling *m*

2531 spring barley
f orge *m* de printemps
e cebada *f* de primavera
i orzo *m* primaverile

p cevada *f* de primavera
d Sommergerste *f*

2532 spring budding
f écussonnage *m* en printemps
e injerto *m* de primavera
i innesto *m* a gemma vegetante
p enxerto *m* de primavera
d Okulation *m* mit treibendem Auge

2533 spring corn; spring grain
f céréales *fpl* de printemps
e cereales *mpl* de verano
i cereali *mpl* primaverili
p cereais *mpl* de verão
d Sommergetreide *n*

2534 spring crop
f récolte *f* de printemps
e cultivo *m* de primavera
i coltura *f* primaverile
p cultura *f* de primavera
d Frühjahrskultur *f*

2535 spring cultivation
f travaux *mpl* de printemps
e labores *fpl* de primavera
i lavori *mpl* primaverili
p cultivos *mpl* de primavera
d Frühjahrsbestellung *f*

*** spring grain → 2533**

2536 spring icing
f givrage *m* printanier
e helada *f* primaveral
i gelata *f* primaverile
p geada *f* primaveril
d Frühlingsaufeis *n*

2537 spring rye
f seigle *m* de printemps
e centeno *m* de primavera
i segale *f* primaverile
p centeio *m* de primavera
d Sommerroggen *m*

*** spring tide → 969**

2538 spring water
f eau *f* de source
e agua *f* de manantial
i acqua *f* di sorgente
p água *f* de fonte
d Quellwasser *n*

2539 spring wheat
f froment *m* de printemps
e trigo *m* de primavera
i frumento *m* primaverile
p trigo *m* de primavera
d Sommerweizen *m*

2540 sprinkler
f arroseur *m*
e aspersor *m*
i irrigatore *m*
p regador *m*
d Regner *m*

2541 sprinkler installations for frost protection
f installations *fpl* d'aspersion antigel
e instalaciones *fpl* para riego por aspersión antiheladas
i impianti *mpl* completi per l'irrigazione a pioggia antigelo
p instalações *fpl* para irrigação por aspersão anti-geadas
d Frostschutzberegnungsanlagen *fpl*

2542 squall
(strong gust of wind, normally accompanied by rain, snow or sleet)
f grain *m*
e turbonada *f*
i groppo *m*
p borrasca *f*
d Bö *f*

*** squall with rain → 2211**

2543 squally weather
f temps *m* à grains
e tiempo *m* chubascoso
i tempo *m* da groppi
p tempo *m* tormentoso
d böiges Wetter *n*

2544 square
f carré *m*
e cuadrado *m*
i quadrato *m*
p quadrado *m*
d Quadrat *n*

2545 square foot
f pied *m* carré
e pie *m* cuadrado
i piede *m* quadrato
p pé *m* quadrado
d Quadratfuß *m*

2546 stability
f stabilité *f*
e estabilidad *f*
i stabilità *f*
p estabilidade *f*
d Stabilität *f*

2547 standard atmosphere
f atmosphère *f* normale
e atmósfera *f* tipo
i aria *f* tipo
p atmosfera *f* modelo
d Normalatmosphäre *f*

2548 standard barometer
f baromètre *m* de contrôle
e barómetro *m* patrón
i barometro *m* campione
p barômetro *m* padrão
d Prüfungsbarometer *n*

2549 standard deviation
f écart-type *m*
e desviación *f* característica
i scarto *m* quadratico medio
p desvio *m* padrão
d Standardabweichung *f*

2550 standard pressure
f pression *f* normale
e presión *f* normal
i pressione *f* tipo
p pressão *f* padrão
d Normaldruck *m*

2551 standard time
f temps *m* legal
e hora *f* civil
i tempo *m* medio locale
p hora *f* normal
d Normalzeit *f*

2552 starlight
f lumière *f* stellaire
e luz *f* de estrellas
i luce *f* stellare
p luz *f* estelar
d Sternlicht *n*

*** state of ground → 2440**

2553 static pressure
f pression *f* statique
e presión *f* estática
i pressione *f* statica
p pressão *f* estática
d statischer Druck *m*

2554 static stability
f stabilité *f* statique
e estabilidad *f* estática
i stabilità *f* statica
p estabilidade *f* estática
d statische Stabilität *f*

2555 stationary air
f air *m* stationnaire
e aire *m* estacionario
i aria *f* stazionaria
p ar *m* estacionário
d stehende Luft *f*

2556 station barometer
f baromètre *m* de station
e barómetro *m* de estación
i barometro *m* da stazione
p barômetro *m* de estação
d Stationsbarometer *n*

2557 statistical analysis
f méthode *f* statistique
e método *m* estadístico
i analisi *f* di statistica
p análise *f* estatística
d mathematische Bearbeitung *f*

2558 statistical test
f test *m* statistique
e prueba *f* estadística
i prova *f* statistica
p prova *f* estatística
d statistischer Test *m*

2559 steadiness
f persistance *f*
e constancia *f*
i persistenza *f*
p persistência *f*
d Beständigkeit *f*

2560 steady rain
f chute *f* de pluie
e lluvia *f* constante
i pioggia *f* costante
p chuva *f* constante
d Landregen *m*

*** steady wind → 621**

*** steam, under ~ → 2859**

2561 steering
f guidage *m*
e acción *f* rectora
i azione *f* direttrice
p guiagem *f*
d Steuern *n*

2562 Stefan-Boltzmann law
f loi *f* de Stefan-Boltzmann
e ley *f* de Stefan-Boltzmann
i legge *f* di Stefan-Boltzmann
p lei *f* de Stefan-Boltzmann
d Stefan-Boltzmannsches Gesetz *n*

2563 steppe; steppe soil
(extensive semi-arid land in southeastern Europe and Asia)
f steppe *f*
e estepa *f*
i steppa *f*
p estepe *f*
d Steppe *f*; Steppenboden *m*

* **steppe soil → 2563**

2564 Stevenson screen
f grille *f* de Stevenson
e pantalla *f* protectora Stevenson
i schermo *m* di Stevenson
p grade *f* de Stevenson
d Stevensonschirm *m*

* **still air → 444**

2565 stimulation of germination
f stimulation *f* de la germination
e estimulación *f* de la germinación
i stimolo *m* della germinazione
p estimulação *f* à germinação
d Keimstimulation *f*

2566 stone desert
f désert *m* rocheux
e desierto *m* rocoso
i deserto *m* roccioso
p deserto *m* rochoso
d Steinwüste *f*

2567 storm; thunderstorm
f tempête *f*; orage *m*
e tempestad *f*; tormenta *f*
i temporale *m*; tempesta *f*
p tormenta *f* ; tempestade *f*
d Sturm *m*; Gewitterbö *f*

2568 storm flood
f grande marée *f*
e gran marea *f*
i grande marea *f*
p grande maré *f*
d Sturmflut *f*

2569 storm region
f région *f* des tempêtes
e región *f* de las tormentas
i regione *f* delle tempeste
p região *f* das tormentas
d Sturmregion *f*

* **storm warning → 1169**

* **stormy** *adj* **→ 3039**

2570 stormy weather
f temps *m* orageux
e tiempo *m* tempestuoso
i tempo *m* burrascoso
p tempo *m* tempestuoso
d stürmisches Wetter *n*

2571 stratification of ice
f stratification *f* de la glace
e estratificación *f* del hielo
i stratificazione *f* del ghiaccio
p estratificação *f* do gelo
d Eisschichtung *f*

2572 stratiform *adj*
f stratiforme
e estratiforme
i stratiforme
p estratiforme
d stratusförmig

2573 stratocumulus
(type of white or grey cloud characterized by stratiform layers)
f strato-cumulus *m*
e estratocumulus *m*
i stratocumulus *m*
p estratocúmulo *m*
d Stratokumulus *m*

2574 stratopause
f stratopause *f*
e estratopausa *f*
i stratopausa *f*
p estratopausa *f*
d Stratopause *f*

2575 stratoscope
f stratoscope *m*
e estratoscopio *m*
i stratoscopio *m*
p estratoscópio *m*
d Stratoskop *n*

2576 stratosphere
f stratosphère *f*
e estratosfera *f*
i stratosfera *f*
p estratosfera *f*
d Stratosphäre *f*

2577 stratosphere aircraft
f avion *m* stratosphérique
e avión *m* estratosférico
i aeroplano *m* estratosferico
p avião *m* estratosférico
d Stratosphärenflugzeug *n*

2578 stratospheric aviation
f aviation *f* stratosphérique
e aviación *f* estratosférica
i aviazione *f* stratosferica
p aviação *f* estratosférica
d stratosphärische Luftfahrt *f*

2579 stratospheric fall-out
f retombée *f* stratosphérique
e depósito *m* estratosférico
i ricaduta *f* stratosferica
p precipitação *f* estratosférica
d Stratosphärenausfall *m*

2580 stratostat
f stratostate *m*
e estratóstato *m*
i stratostrato *m*
p estratóstato *m*
d Stratostat *n*

* **stratum of clouds → 551**

2581 stratus
f stratus *m*
e stratus *m*
i strato *m*
p estrato *m*
d Stratus *m*

2582 stratus cloud
f nuage *m* stratus
e nube *f* stratus
i nuvola *f* strato
p nuvem *f* stratus
d Schichtwolke *f*

2583 stray earth currents
f courants *mpl* vagabonds
e corrientes *fpl* vagabundas
i correnti *fpl* vagabonde
p correntes *fpl* vagabundas
d vagabundierende Ströme *mpl*

* **stream → 698**

* **stream current → 1959**

2584 stream ice
f glace *f* des rivières
e hielo *m* de los ríos
i ghiaccio *m* di fiume
p gelo *m* dos rios
d Bacheis *n*

2585 strength of ice
f résistance *f* de la glace
e resistencia *f* del hielo
i resistenza *f* del ghiaccio
p resistência *f* do gelo
d Eisfestigkeit *f*

* **strength of snow → 2272**

2586 stress of weather
f temps *m* forcé
e fuerza *f* del tiempo
i violenza *f* di tempo
p tensão *f* do tempo
d Unwetter *n*

2587 strip cropping; strip farming
f culture *f* en bandes
e cultivo *m* en fajas; cultivo *m* en franjas
i coltivazione *f* a strisce
p cultivo *m* em faixas
d Streifenkultur *f*

* **strip farming → 2587**

2588 strong *adj*
f fort
e fuerte
i forte
p forte
d stark

* **strong breeze → 2591**

2589 strong gale
f fort coup *m* de vent
e viento *m* muy duro
i burrasca *f* forte
p vento *m* duro; ventania *f* forte
d Sturm *m*

2590 strongly agitated air
f air *m* fortement agité
e aire *m* fuertemente agitado
i aria *f* fortemente agitata
p ar *m* fortemente agitado
d stark bewegte Luft *f*

2591 strong wind; strong breeze; fresh gale
f vent *m* fort; vent *m* frais
e viento *m* fuerte; brisa *f* fuerte
i vento *m* forte; brezza *f* intensa
p vento *m* forte; brisa *f* forte
d starker Wind *m*; stürmischer Wind *m*

2592 structural glaciology
f glaciologie *f* structurale
e glaciología *f* estructural
i glaciologia *f* strutturale
p glaciologia *f* estrutural
d strukturelle Glaziologie *f*

2593 structure of ice
f structure *f* de la glace
e estructura *f* del hielo
i struttura *f* del ghiaccio
p estrutura *f* do gelo
d Eisstruktur *f*

2594 subauroral belt
f ceinture *f* subaurorale
e cinturón *m* subauroral
i fascia *f* subaurorale
p cinturão *m* subauroral
d Gürtel *m* südlich der Polarlichtzone

2595 subcloud layer
f couche *f* de convection au dessous des nuages
e capa *f* estable de nubes de convección situada debajo de las nubes
i strato *m* sotto le nubi
p camada *f* abaixo das nuvens
d Schicht *f* unter den Wolken

2596 subglacial ablation
f ablation *f* sous-glaciaire
e ablación *f* subglacial
i ablazione *f* subglaciale
p ablação *f* subglacial
d subglaziale Ablation *f*

2597 subglacial channel
f canal *m* sous-glaciaire
e canal *m* subglacial
i canale *m* subglaciale
p canal *m* subglacial
d subglaziales Gerinne *n*

2598 subglacial lake
f lac *m* sous-glaciaire
e lago *m* subglacial
i lago *m* subglaciale
p lago *m* subglacial
d subglazialer See *m*

2599 subglacial water
f eau *f* sous-glaciaire
e agua *f* subglacial
i acqua *f* subglaciale
p água *f* subglacial
d subglaziales Wasser *n*

2600 sublimation
f sublimation *f*
e sublimación *f*
i sublimazione *f*
p sublimação *f*
d Sublimation *f*

2601 submarine earthquake
f tremblement *m* de terre sous-marin
e terremoto *m* submarino
i terremoto *m* sottomarino
p maremoto *m*
d Seebeben *n*

2602 submarine fog bell
f cloche *f* de brume sous-marine
e campaña *f* de bruma submarina
i campana *f* sottomarina da nebbia
p sinal *m* acústico submarino de nevoeiro
d Unterwassernebelglocke *f*

2603 submarine fog signal
f signal *m* de brume sous-marin
e señal *f* de niebla submarina
i segnale *m* di nebbia sottomarino
p sinal *f* de neblina submarina
d Unterwassernebelsignal *n*

2604 submarine volcano
f volcan *m* sous-marin
e volcán *m* submarino
i vulcano *m* sottomarino
p vulcão *m* submarino
d untermeerischer Vulkan *m*

2605 submerged ice
f glace *f* inondée
e hielo *m* hundido
i ghiaccio *m* sommerso
p gelo *m* submerso
d untergetauchtes Eis *n*

2606 subsidence
f subsidence *f*
e subsidencia *f*
i subsidenza *f*
p abatimento *m*
d Absinken *n*

2607 subsurface irrigation
f irrigation *f* de subsurface
e riego *m* subterráneo
i subirrigazione *f*
p rega *f* subterrânea
d Untergrundwasserung *f*

2608 subtropical air
f air *m* subtropical
e aire *m* subtropical
i aria *f* subtropicale
p ar *m* subtropical
d subtropische Luft *f*

2609 subtropical environment
f zone *f* subtropicale
e zona *f* subtropical
i zona *f* subtropicale

p ambiente *m* subtropical
d Subtropenzone *f*

2610 subtropical high-pressure belt
f région *f* des calmes subtropicaux
e zona *f* de calmas subtropicales
i cintura *f* subtropicale di alta pressione
p zona *f* de calmas subtropicais
d subtropischer Stillengürtel *m*

2611 subtropic region
f région *f* subtropicale
e región *f* subtropical
i regione *f* subtropicale
p região *f* subtropical
d subtropisches Gebiet *n*

2612 suburban farming
f agriculture *f* de banlieue
e agricultura *f* suburbana
i agricoltura *f* suburbana
p agricultura *f* suburbana
d stadtnahe Landwirtschaft *f*

2613 sudden change of temperature
f variation *f* brusque de température
e variación *f* brusca de temperatura
i variazione *f* brusca di temperatura
p mudança *f* brusca de temperatura
d plötzliche Änderung *f* der Temperatur

2614 sudden ionospheric disturbance
f perturbation *f* ionosphérique à début brusque
e perturbación *f* ionosférica súbita
i disturbo *m* ionosferico improvviso
p perturbação *f* ionosférica brusca
d plötzliche Ionosphärenstörung *f*

2615 sudden shift of the wind
f saute *f* de vent
e salto *m* de viento
i salto *m* di vento
p salto *m* do vento
d Umspringen *n* des Windes

2616 sulphur dioxide
f bioxyde de soufre
e dióxido *m* de azufre
i biossido *m* di zolfo
p dióxido *m* de enxôfre
d Schwefeldioxid *n*

2617 sulphuric acid
f acide *m* sulfurique
e ácido *m* sulfúrico
i acido *m* solforico
p ácido *m* sulfúrico
d Schwefelsäure *f*

2618 sultry *adj*
f étouffant
e sofocante
i afoso; soffocante
p abafador
d schwül

2619 sultry weather; muggy weather; oppressive weather
f temps *m* lourd et humide
e tiempo *m* sofocante; tiempo *m* pesado y húmedo
i tempo *m* soffocante; tempo *m* pesante e umido
p tempo *m* sufocante; tempo *m* carregado e úmido; mormaço *m*
d schwüles Wetter *n*; drückendes Wetter *n*

2620 summer
f été *m*
e verano *m*
i estate *f*
p verão *m*; estio *m*
d Sommer *m*

2621 summer annual
f annuelle *f* d'été
e anual *m* estival
i annuale *f* estiva
p planta *f* anual de verão
d Sommerannuell *f*

2622 summer catch crop
f culture *f* dérobée d'été
e cultivo *m* intermedio de verano
i coltura *f* intercalare d'estiva
p cultura *f* intermediária de verão
d Sommerzwischenfrucht *f*

2623 summer crop
f récolte *f* d'été
e cosecha *f* de verano
i coltura *f* d'estiva
p cultura *f* de verão
d Sommerkultur *f*

2624 summer cutting; summer pruning
f taille *f* estivale
e poda *f* estival
i potatura *f* estiva
p poda *f* de verão
d Sommerschnitt *m*

2625 summer diarrhea
f diarrhée *f* d'été
e diarrea *f* de verano
i diarrea *f* estiva

p diarréia *f* de verão
d Sommerdiarrhöe *f*

2626 summer dyke
f digue *f* d'été
e dique *m* de estiaje
i argine *m* golenale
p dique *m* de estiagem
d Sommerdeich *m*

2627 summering; summer pasturing
f estivage *m*; estive *f*
e agostadero *m*; pasto *m* de estío
i estivamento *m*
p pasto *m* de verão
d Sömmerung *f*

2628 summer lightning
f éclair *m* d'été
e relámpago *m* de verano
i lampo *m* di estivo
p trovoadas *fpl* de verão
d Wetterleuchten *n*

2629 summer monsoon
f mousson *f* d'été
e monzón *m* de verano
i monsone *m* d'estate
p monção *f* estival
d Sommermonsun *m*

2630 summer navigation
f navigation *f* d'été
e navegación *f* de verano
i navigazione *f* estiva
p navegação *f* de verão
d Sommerfahrt *f*

* **summer pasturing → 2627**

* **summer pruning → 2624**

2631 summer sea marks
f amers *mpl* d'été
e marcas *fpl* de estío
i segnali *mpl* marittimi d'estate
p marcas *fpl* de estío
d Sommerseezeichen *npl*

2632 summer solstice
f solstice *m* d'été
e solsticio *m* de estío
i solstizio *m* d'estate
p solstício *m* de verão
d Sommersonnenwende *f*

2633 summer spraying
f pulvérisation *f* d'été
e pulverización *f* de verano
i trattamento *m* estivo
p pulverização *f* de verão
d Sommerspritzung *f*

2634 summer surface
f surface *f* estivale
e superficie *f* de estío
i superficie *f* estivale
p superfície *f* de verão
d Sommeroberfläche *f*

2635 summertime
f heure *f* d'été
e hora *f* de verano
i ora *f* estiva
p hora *f* de verão
d Sommerzeit *f*

2636 summer truffle
f truffe *f* d'été
e trufa *f* de verano
i tartufo *m* d'estate
p tubera *f* de verão
d Sommertrüffel *f*

2637 sum of precipitation
f somme *f* des précipitations
e suma *f* de las precipitaciones
i somma *f* delle precipitazioni
p soma *f* das precipitações
d Niederschlagssumme *f*

2638 sun
f soleil *m*
e sol *m*
i sole *m*
p sol *m*
d Sonne *f*

2639 sunbath
f bain *m* de soleil
e baño *m* de sol
i bagno *m* di sole
p banho *m* de sol
d Sonnenbad *n*

2640 sunbathing area; sun terrace
f solarium *m*
e zona *f* de reposo en el sol
i zona *f* per bagni di sole
p zona *f* de banhos-de-sol; solário *m*
d Liegewiese *f*

2641 sundial
f cadran *m* solaire
e cuadrante *m* solar
i orologio *m* solare
p relógio *m* de sol
d Sonnenuhr *f*

2642 sunlight
f lumière *f* solaire
e luz *f* solar
i luce *f* solare
p luz *f* solar
d Sonnenlicht *n*

2643 sun pillar
f colonne *f* solaire
e columna *f* solar
i colonna *f* solare
p coluna *f* solar
d Sonnensäule *f*

2644 sunrise
f lever *m* du soleil
e salida *f* del sol
i sorgere *m* del sole
p nascer-do-sol *m*
d Sonnenaufgang *m*

2645 sunrise and sunset colours
f couleurs *fpl* du levant et du couchant
e coloraciones *fpl* del orto y ocaso
i colori *mpl* del sorgere e del tramonto del sole
p colorações *fpl* ao nascer e pôr do sol
d Dämmerungsfarben *fpl*

2646 sun's azimuth
f azimut *m* du soleil
e azimut *m* del sol
i azimut *m* del sole
p azimute *m* do sol
d Sonnenazimut *m*

2647 sun scald
f brûlure *f* de soleil
e quemadura *f* del sol
i scottatura *f* solare
p escaldão *m* solar
d Sonnenbrand *m*

2648 sunset
f coucher *m* du soleil
e puesta *f* del sol
i tramonto *m* del sole
p pôr-do-sol *m*
d Sonnenuntergang *m*

2649 sunset colours; sunset glow
f couleurs *fpl* du couchant
e colores *mpl* del crepúsculo
i rosso *m* del tramonto; colori *mpl* del tramonto
p cores *fpl* do pôr-do-sol
d Abendröte *f*; Abendrot *n*

* **sunset glow → 2649**

2650 sunshade
f pare-soleil *m*
e guardasol *m*
i parasole *m*
p pára-sol *m*
d Sonnenschirm *m*

2651 sunshine
f insolation *f*
e luz *f* de sol
i splendore *m* del sole
p brilho *m* do sol
d Sonnenschein *m*

* **sunshine recorder → 1352**

2652 sunspot
f tache *f* solaire
e mácula *f* del sol
i macchia *f* del sole
p mancha *f* solar
d Sonnenfleck *m*

2653 sunspot number
f nombre *m* des taches solaires
e número *m* de manchas solares
i numero *m* delle macchie solari
p número *m* de manchas solares
d Sonnenfleckenzahl *f*

* **sunstroke → 2479**

* **sun terrace → 2640**

2654 superadiabatic lapse rate
f gradient *m* super-adiabatique
e gradiente *m* superadiabático vertical de temperatura
i decremento *m* di temperatura superadiabatico
p gradiente *m* super-adiabático
d superadiabatische Maßveränderung *f*

2655 superadiabatic layer
f couche *f* super-adiabatique
e capa *f* superadiabática
i strato *m* superadiabatico
p camada *f* super-adiabática
d superadiabatische Schicht *f*

2656 supercooled water
f eau *f* surfondue
e agua *f* sobrefundida
i acqua *f* sopraraffreddata
p água *f* superfundida
d untergekühltes Wasser *n*

2657 supercooling
f surfusion *f*
e sobrefusión *f*
i sopraffusione *f*
p sobrefusão *f*
d Unterkühlung *f*

2658 supersaturation
f sursaturation *f*
e sobresaturación *f*
i soprasaturazione *f*
p sobressaturação *f*
d Übersättigung *f*

2659 supply current
f courant *m* de compensation
e corriente *f* de alimentación
i corrente *f* di alimentazione
p corrente *f* de compensação
d Speisestrom *m*

2660 suprasphere
f suprasphère *f*
e suprasfera *f*
i suprasfera *f*
p suprasfera *f*
d Suprasphäre *f*

2661 surface contamination meter
f contaminamètre *m* surfacique
e medidor *m* de contaminación superficial
i contaminametro *m* di superficie
p medidor *m* de contaminação da superfície
d Gerät *n* zur Bestimmung der Oberflächenkontamination

2662 surface of discontinuity
f surface *f* de discontinuité
e superficie *f* de discontinuidad
i superficie *f* di discontinuità
p superfície *f* de descontinuidade
d Diskontinuitätsfläche *f*

2663 surface of subsidence
f surface *f* de subsidence
e superficie *f* de subsidencia
i superficie *f* di subsidenza
p superfície *f* de abatimento
d Abgleitfläche *f*

2664 surge
f augmentation *f* générale de la pression atmosphérique
e cambio *m* general de la presión barométrica
i aumento *m* della pressione atmosferica
p mudança *f* brusca de pressão atmosférica
d allgemeines Ansteigen *n* des Luftdruckes

* **swamp → 1754**

2665 swampy coast
f côte *f* marécageuse
e costa *f* pantanosa
i costa *f* paludosa
p costa *f* pantanosa
d sumpfige Küste *f*

2666 swash
f vague *f* d'arrivée
e marisma *f*
i getto *m* di riva
p restinga *f*
d Riff *n*; Sandbank *f*

2667 sweat
f sueur *f*
e sudor *m*
i sudore *m*
p suor *m*
d Schweiß *m*

2668 sweat *v*
f suer
e sudar
i sudare
p suar
d schwitzen

2669 swell
f houle *f*
e mar *m* de leva
i onda *f*
p crescida *f*
d Dünung *f*

2670 synodic period
f période *f* synodique
e período *m* sinódico
i periodo *m* sinodico
p período *m* sinódico
d synodischer Periode *f*

2671 synoptic chart
f carte *f* synoptique du temps
e mapa *m* sinóptico del tiempo
i carta *f* sinottica del tempo
p mapa *m* sinóptico do tempo
d synoptische Wetterkarte *f*

2672 synoptic meteorology
f météorologie *f* synoptique
e meteorología *f* sinóptica
i meteorologia *f* sinottica

p meteorologia *f* sinóptica
d synoptische Meteorologie *f*

2673 synoptic report
f bulletin *m* synoptique
e reportaje *m* sinóptico
i relazione *f* sinottica
p boletim *m* sinóptico
d synoptische Meldung *f*

2674 syzygy
(the position of a celestial object's orbit in conjunction with, or in opposition to, the sun)
f syzygie *f*
e sicigia *f*
i sizigia *f*
p sizígia *f*
d Syzygie *f*

T

2675 tabular iceberg
f iceberg *m* tabulaire
e iceberg *m* tabular
i iceberg *m* tabulare
p icebergue *m* tabular
d Tafeleisberg *m*

2676 taiga
(Siberian forest vegetation)
f taïga *f*
e taiga *f*
i taiga *f*
p taiga *f*
d Taiga *f*

2677 tail wind; fair wind
f vent *m* d'arrière
e viento *m* de cola
i vento *m* di cola
p vento *m* pela cauda
d Rückenwind *m*

2678 take *v* **the sun's altitude**
f prendre la hauteur du soleil
e tomar la altura del sol
i prendere l'altezza del sole
p tomar a altura do sol
d die Sonnenhöhe messen

2679 tectonic earthquake
f tremblement *m* volcanique
e terremoto *m* tectónico
i terremoto *m* tettonico
p terremoto *m* tectônico
d tektonisches Beben *n*

2680 tectonic forces
f forces *fpl* tectoniques
e fuerzas *fpl* tectónicas
i forze *fpl* tettoniche
p forças *fpl* tectônicas
d tektonische Kräfte *fpl*

2681 tectonics of ice
f tectonique *f* de la glace
e tectónica *f* del hielo
i tettonico *m* del ghiaccio
p tectônica *f* do gelo
d Eistektonik *f*

2682 tehuantpecer
(violent north or north-northeast wind around the Gulf of Tehuantepec (Mexico))
f tehuantpecer *m*
e viento *m* de Tehuantpec
i tehuantpecer *m*
p tehuantpecer *m*
d Tehuantpecerwind *m*

2683 telemetry antenna
f antenne *f* télémétrique
e antena *f* telemétrica
i antenna *f* telemetrica
p antena *f* telemétrica
d Telemetrieantenne *f*

2684 television camera
f caméra *f* de télévision
e cámara *f* de televisión
i telecamera *f*
p câmara *f* de televisão
d Fernsehkamera *f*

2685 temperate climate
f climat *m* tempéré
e clima *m* templado
i clima *m* temperato
p clima *m* temperado
d gemäßigtes Klima *n*

2686 temperate glacier; isothermal glacier
f glacier *m* tempéré
e glaciar *m* temperado
i ghiacciaio *m* temperato
p geleira *f* temperada
d temperierter Gletscher *m*

2687 temperate glasshouse
f serre *f* tempérée
e invernadero *m* templado
i serra *f* temperata
p clima *m* de estufa temperado
d temperiertes Gewächshaus *n*

2688 temperate zone
f zone *f* tempérée
e zona *f* templada
i zona *f* temperata
p zona *f* temperada
d gemäßigte Zone *f*

2689 temperature
f température *f*
e temperatura *f*
i temperatura *f*
p temperatura *f*
d Temperatur *f*

* **temperature drop → 739**

* **temperature gauge → 2731**

2690 temperature gradient
f gradient *m* de température
e gradiente *m* de temperatura
i gradiente *m* di temperatura
p gradiente *m* de temperatura
d Temperaturgradient *m*

2691 temperature gradient stress
f tension *f* due aux chutes de température
e esfuerzo *m* debido al gradiente de temperatura
i sollecitazione *f* dovuta a gradiente di temperatura
p tensão *f* devida às variações de temperatura
d Temperaturgradientspannung *f*

2692 temperature graph
f courbe *f* des températures
e gráfica *f* de las temperaturas
i grafico *m* delle temperature
p gráfico *m* das temperaturas
d Temperaturlinie *f*

* **temperature increase → 2278**

* **temperature in the shade → 2362**

2693 temperature inversion
f inversion *f* de température
e inversión *f* de temperatura
i inversione *f* di temperatura
p inversão *f* de temperatura
d Temperaturinversion *f*

* **temperature range → 2882**

2694 temperature sensor
f sonde *f* thermométrique
e sensor *m* térmico
i sensore *m* della temperatura
p sensor *m* de temperatura
d Temperatursensor *m*

2695 temperature sum; heat sum
f somme *f* thermique
e suma *f* de temperaturas
i somma *f* termica
p soma *f* de temperaturas
d Temperatursumme *f*

2696 temperature survey
f contrôle *m* thermique
e control *m* térmico
i controllo *m* termico
p registro *m* de temperatura
d Wärmekontrolle *f*

2697 tensile strength of snow
f résistance *f* à la traction de la neige
e resistencia *f* a la tracción de la nieve
i resistenza *f* alla trazione della neve
p resistência *f* à tração da neve
d Ausspannungsfestigkeit *f* des Schnees

2698 terrestrial *adj*
f terrestre
e terrestre
i terrestre
p terrestre
d erdisch

2699 terrestrial equator
f équateur *m* terrestre
e ecuador *m* terrestre
i equatore *m* terrestre
p equador *m* terrestre
d Erdäquator *m*

2700 terrestrial globe
f globe *m* terrestre
e globo *m* terrestre
i globo *m* terrestre
p globo *m* terrestre
d Erdkugel *f*

2701 terrestrial magnetic axis
f axe *m* magnétique terrestre
e eje *m* magnético terrestre
i asse *m* magnetico terrestre
p eixo *m* magnético terrestre
d geomagnetische Achse *f*

* **terrestrial magnetism → 886**

2702 terrestrial meridian
f méridien *m* terrestre
e meridiano *m* terrestre
i meridiano *m* terrestre
p meridiano *m* terrestre
d Ortsmeridian *m*; Erdmeridian *m*

2703 terrestrial radiation
f rayonnement *m* terrestre
e radiación *f* terrestre
i radiazione *f* terrestre
p radiação *f* terrestre
d Erdstrahlung *f*

2704 texture of ice
f texture *f* de la glace
e textura *f* del hielo
i tessitura *f* del ghiaccio
p textura *f* do gelo
d Eistextur *f*

2705 thalassotherapy
f thalassothérapie *f*
e talasoterapia *f*
i talassoterapia *f*
p talassoterapia *f*
d Behandlung *f* mit Meerwasser

2706 thaw
f dégel *m*
e deshielo *m*
i disgelo *m*
p degelo *m*
d Tauwetter *n*

*** thawing point → 790**

2707 thaw weather
f temps *m* de dégel
e tiempo *m* de deshielo
i tempo *m* di disgelo
p tempo *m* de degelo
d Tauwetter *n*

2708 theodolite
(instrument in meteorology used to observe the motion of a pilot balloon)
f théodolite *m*
e teodolito *m*
i teodolite *m*
p teodolito *m*
d Theodolit *m*

*** thermal depression → 2715**

2709 thermal efficiency
f efficacité *f* thermique
e rendimiento *m* térmico
i rendimento *m* termico
p rendimento *m* térmico
d thermischer Wirkungsgrad *m*

2710 thermal equator
f équateur *m* thermique
e ecuador *m* térmico
i equatore *m* termico
p equador *m* térmico
d Wärmeäquator *m*

2711 thermal gradient
f gradient *m* thermique
e gradiente *m* térmico
i gradiente *m* termico
p gradiente *m* térmico
d thermischer Gradient *m*

2712 thermal influence of the moon
f influence *f* thermique de la lune
e influencia *f* térmica de la luna
i influenza *f* termica della luna
p influência *f* térmica da lua
d Wärmewirkung *f* des Mondes

2713 thermal insulation
f insolation *f* thermique
e aislamiento *m* térmico
i isolazione *f* termica
p isolamento *m* térmico
d Wärmeschutz *m*

2714 thermal lift
f ascendance *f* thermique
e ascendencia *f* térmica
i corrente *f* ascendente termica
p ascendência *f* térmica
d thermischer Aufwind *m*

2715 thermal low; thermal depression
f dépression *f* thermique
e depresión *f* térmica
i bassa *f* termica
p depressão *f* térmica
d Wärmetief *n*

2716 thermal metamorphism
f métamorphisme *m* thermique
e metamorfismo *m* térmico
i metamorfismo *m* termico
p metamorfismo *m* térmico
d Thermometamorphose *f*

2717 thermal metamorphism of ice
f thermométamorphisme *m* de la glace
e metamorfismo *m* térmico del hielo
i metamorfismo *m* termico del ghiaccio
p metamorfismo *m* térmico do gelo
d Thermometamorphose *f* des Eises

2718 thermal screen
f écran *m* thermique
e pantalla *f* térmica
i schermo *m* termico
p cortina *f* térmica
d Energieschirm *m*

2719 thermal spring
f eau *f* thermale
e fuente *f* termal
i sorgente *f* termale
p fonte *f* termal
d Thermalquelle *f*

2720 thermal stability
f stabilité *f* thermique
e termoestabilidad *f*
i stabilità *f* termica
p estabilidade *f* térmica
d Wärmebeständigkeit *f*

2721 thermal stratification
f stratification *f* thermique
e estratificación *f* térmica
i stratificazione *f* termica
p estratificação *f* térmica
d Temperaturschichtung *f*

2722 thermal streets
f émanations *fpl* thermiques d'air ascendant
e gamas *fpl* térmicas con movimiento vertical
i mastri *mpl* termici
p faixas *fpl* térmicas de movimento
d Wärmestraßen *fpl*

2723 thermal unit
f thermie *f*
e unidad *f* térmica
i unità *f* termica
p unidade *f* térmica
d Wärmeeinheit *f*

2724 thermal wind
f vent *m* thermique
e viento *m* térmico
i vento *m* termico
p vento *m* térmico
d thermischer Wind *m*

2725 thermistor
f thermistor *m*
e termistor *m*
i termistore *m*
p termistor *m*
d Thermistor *m*

2726 thermocouple
f thermocouple *m*
e par *m* termoeléctrico
i termocoppia *f*
p binário *m* termo-elétrico
d Thermoelement *n*

2727 thermodynamics
f thermodynamique *f*
e termodinámica *f*
i termodinamica *f*
p termodinâmica *f*
d thermodynamische Wärmelehre *f*

2728 thermogram
f thermogramme *m*
e termograma *m*
i termogramma *m*
p termograma *m*
d Thermogramm *n*

2729 thermograph; recording thermometer
f thermographe *m*
e termógrafo *m*
i termografo *m*
p termógrafo *m*
d Thermograph *m*

2730 thermography
f thermographie *f*
e termografía *f*
i termografia *f*
p termografia *f*
d Thermographie *f*

2731 thermometer; temperature gauge
f thermomètre *m*; appareil *m* de mesure de la température
e termómetro *m*; medidor *m* de temperatura
i termometro *m*; apparecchio *m* per misurare la temperatura
p termômetro *m*
d Thermometer *n*

2732 thermometer screen
f boîte *f* de thermomètre
e pantalla *m* del termómetro
i pannello *m* del termometro
p grade *f* termométrica
d Thermometergehäuse *n*

2733 thermometric *adj*
f thermométrique
e termométrico
i termometrico
p termométrico
d thermometrisch

2734 thermometric scale
f échelle *f* thermométrique
e escala *f* termométrica
i scala *f* termometrica
p escala *f* termométrica
d Thermometerskala *f*

2735 thermometric temperature
f température *f* thermométrique
e temperatura *f* termométrica
i temperatura *f* termometrica
p temperatura *f* termométrica
d thermometrische Temperatur *f*

2736 thermometry
f thermométrie *f*
e termometría *f*
i termometria *f*
p termometria *f*
d Wärmemessung *f*

2737 thermoscope
f thermoscope *m*
e termoscopio *m*

i termoscopio *m*
p termoscópio *m*
d Thermoskop *n*

2738 thermosphere
f thermosphère *f*
e termosfera *f*
i termosfera *f*
p termosfera *f*
d Thermosphäre *f*

2739 thick fog
f brouillard *m* épais
e niebla *f* densa
i nebbia *f* fitta
p nevoeiro *m* denso
d dichter Nebel *m*

* **thick squall → 2211**

2740 thick weather
f temps *m* épais
e tiempo *m* espeso
i tempo *m* denso
p tempo *m* fechado
d dickes Wetter *n*

2741 thoron
(radioactive gaseous decay product formed by disintegration of thorium)
f thoron *m*
e torón *m*
i toron *m*
p tóron *m*
d Thoron *n*

2742 threatening sky
f ciel *m* orageux
e cielo *m* amenazador
i tempo *m* minaccioso
p céu *m* ameaçador
d gewitterdrohender Himmel *m*

* **three-body collision → 2797**

2743 thunder
f tonnerre *m*
e trueno *m*
i tuono *m*
p trovão *m*
d Donner *m*

* **thunderbolt → 1644**

2744 thundercloud
f nuage *m* orageux
e nube *f* de tormenta
i nuvola *f* temporalesca
p nuvem *f* de trovoada
d Gewitterwolke *f*

2745 thunderhead
f cœur *m* de l'orage
e centro *m* de tormenta
i centro *m* del temporale
p cabeça *f* de trovoada; centro *m* de tempestade
d Gewitterkopf *m*

* **thunderstorm → 2567**

2746 thunderstorm cell
(local disturbance not associated with a front)
f orage *m* local
e sistema *m* de baja presión
i temporale *m* locale
p tempestade *f* local
d lokales Gewitter *n*

2747 thunderstorm high
f anticyclone *m* dû à l'orage
e anticiclón *m* temporal
i anticiclone *m* temporalesco
p anticiclone *m* temporal
d Gewitterhoch *n*

2748 thundery depressions
f orages *mpl* d'été
e depresiones *fpl* tormentosas
i depressioni *fpl* temporaleschi
p tempestades *fpl* locais de verão
d Sommergewitter *npl*

2749 tidal *adj*
f relatif aux marées
e relativo a las mareas
i relativo alle maree
p relativo às marés
d die Gezeiten betreffend

* **tidal amplitude → 2759**

2750 tidal breeze
f brise *f* de marée
e brisa *f* de marea
i brezza *f* di marea
p brisa *f* de maré
d Tidenbrise *f*

2751 tidal chart
f carte *f* des marées
e carta *f* de las mareas
i carta *f* delle maree
p carta *f* das marés
d Gezeitenkarte *f*

2752 tidal constants
f constantes *fpl* de marée
e constantes *fpl* de marea
i costanti *fpl* armoniche
p constantes *fpl* das marés
d Gezeitengrundwerte *mpl*

2753 tidal current; tidal river
f flot *m* de marée
e corriente *f* de la marea
i corrente *f* di marea
p corrente *f* de maré
d Gezeitenströmung *f*

2754 tidal datum
f niveau *m* de référence des marées
e nivel *m* de referencia de las mareas
i livello *m* di riferimento delle maree
p plano *m* de referência das marés
d Gezeitennull *f*

2755 tidal differences
f différences *fpl* marégraphiques
e diferencias *fpl* de marea
i differenze *fpl* di marea
p diferenças *fpl* de marés
d Gezeitenunterschiede *mpl*

2756 tidal glacier
f glacier *m* soumis à la marée
e glaciar *m* del tipo ártico
i ghiacciaio *m* di tipo artico
p glaciar *m* do tipo ártico
d Gezeitengletscher *m*

2757 tidal marsh
f terres *fpl* intercotidales
e tierras *fpl* bañadas por la marea
i sedimento *m* costiero recente
p pântano *m* de maré
d Seemarsch *f*

2758 tidal oscillations
f oscillations *fpl* de marée
e oscilaciones *fpl* de mareas
i oscillazioni *fpl* di marea
p oscilações *fpl* das marés
d Gezeitenschwingungen *fpl*

2759 tidal range; tidal amplitude
f amplitude *f* de la maré
e amplitud *f* de la marea
i ampiezza *f* della marea
p amplitude *f* da maré
d Tidenhub *m*

*** tidal river → 2753**

2760 tidal signal
f signal *m* de marée
e señal *f* de marea
i segnale *m* di marea
p sinal *m* de maré
d Gezeitensignal *n*

2761 tidal wave
f raz *m* de marée
e ola *f* de marea
i onda *f* di marea
p onda *f* de maré
d Flutwelle *f*

2762 tide
f marée *f*
e marea *f*
i marea *f*
p maré *f*
d Gezeiten *fpl*; Tide *f*

2763 tide crack
f crevasse *f* de marée
e crujido *m* de marea
i frattura *f* di marea
p rachadura *f* de maré
d Gezeitenrinne *f*

*** tide gauge → 1747**

2764 tide pole
f échelle *f* de marée
e escala *f* de mareas
i asta *f* mareometrica
p escala *f* de marés
d Lattenpegel *m*

2765 tide-predicting machine
f machine *f* à prédire les marées
e máquina *f* de predicción de mareas
i macchina *f* per la previsione delle maree
p máquina *f* de prognosticar marés
d Gezeitenrechenmaschine *f*

2766 tide way
f lit *m* de marée
e hilero *m* de corriente de marea
i canale *m* di marea
p canal *m* de maré
d Stromstrich *m*

*** tight → 2200**

2767 tillage programme
f plan *m* de culture
e plan *m* de utilización del suelo
i piano *m* di coltivazione
p plano *m* de cultura
d Feldbestellungsplan *m*

2768 time
f temps *m*
e tiempo *m*
i tempo *m*
p tempo *m*
d Zeit *f*

2769 time azimuth
f azimut *m* de temps
e azimut *m* de tiempo
i azimut *m* di tempo
p azimute *m* de tempo
d Zeitazimut *m*

2770 time constant
f constante *f* de temps
e constante *f* de tiempo
i costante *f* di tempo
p constante *f* de tempo
d Zeitkonstante *f*

2771 time of high water
f heure *f* de la pleine mer
e hora *f* de la pleamar
i ora *f* dell'alta marea
p hora *f* da pleamar
d Hochwasserzeit *f*

2772 time of low water
f heure *f* de la marée basse
e hora *f* de la bajamar
i ora *f* della barea
p hora *f* da baixamar
d Niedrigwasserzeit *f*

2773 time of observation
f temps *m* observé
e tiempo *m* de observación
i tempo *m* d'osservazione
p tempo *m* de observação
d Beobachtungszeit *f*

2774 time of the sun's culmination
f temps *m* de la culmination du soleil
e tiempo *m* de la culminación del sol
i tempo *m* della culminazione del sole
p tempo *m* de culminação do sol
d Kulminationszeit *f* der Sonne

2775 time series
f série *f* chronologique
e serie *f* cronológica
i serie *f* storica
p série *f* cronológica
d Zeitreihe *f*

2776 today
f aujourd'hui
e hoy
i oggi
p hoje
d heute

2777 ton
f tonne *f*
e tonelada *f*
i tonnellata *f*
p tonelada *f*
d Tonne *f*

*** top of the atmosphere → 480**

*** tornado → 3001**

*** torrential rain → 2211**

2778 torrid zone
f zone *f* torride
e zona *f* tórrida
i zona *f* torrida
p zona *f* tórrida
d heiße Zone *f*

2779 total eclipse of the moon
f éclipse *f* totale de la lune
e eclipse *m* total de la luna
i eclisse *f* totale di luna
p eclipse *m* total da lua
d totale Mondfinsternis *f*

2780 total precipitation
f grosse précipitation *f*
e precipitación *f* total
i precipitazione *f* totale
p precipitação *f* total
d Bruttoniederschlag *m*

2781 total pressure
f pression *f* totale
e presión *f* total
i pressione *f* totale
p pressão *f* total
d Totaldruck *m*

2782 towering seas
f lames *fpl* très élevées
e olas *fpl* mucho elevadas
i onde *fpl* molto elevate
p ondas *fpl* muito elevadas
d auftürmende Wellen *fpl*

2783 trace of rainfall
f trace *f* de pluie
e lluvia *f* inapreciable
i traccia *f* della pioggia
p vestígio *m* de chuva
d Spur *f*; unmeßbarer Niederschlag *m*

2784 trade drift
f courant *m* de vents alizés
e corriente *f* de los alisios
i corrente *f* dagli alisei
p corrente *f* de ventos alísios
d Passatdrift *f*

* **trades → 2785**

2785 trade winds; trades
f vents *mpl* alizés
e vientos *mpl* alisios
i venti *mpl* alisei
p ventos *mpl* alísios
d Passatwinde *mpl*

2786 trajectory
f trajectoire *f*
e trayectoria *f*
i traiettoria *f*
p trajetória *f*
d Trajektorie *f*

2787 tramontana
(cold wind from the north or northeast of Italia and Corsica across the Adriatic)
f tramontane *f*
e tramontana *f*
i tramontana *f*
p tramontana *f*
d Tramontana *f*

2788 transition zone
f zone *f* de transition
e zona *f* de transición
i zona *f* di transizione
p zona *f* de transição
d Zwischenschicht *f*

2789 transmission coefficient
f coefficient *m* de transmission
e coeficiente *m* de transmisión
i fattore *m* di trasmissione
p coeficiente *m* de transmissão
d Durchlässigkeitsfaktor *m*

2790 transmitter
f émetteur *m*
e emisor *m*; transmisor *m*
i trasmettitore *m*
p transmissor *m*
d Sender *m*

* **transmitter aerial → 2791**

2791 transmitting aerial; transmitter aerial
f antenne *f* émettrice
e antena *f* transmisora
i antenna *f* emittente
p antena *f* transmissora
d Sendeantenne *f*

2792 transparency
f transparence *f*
e transparencia *f*
i trasparenza *f*
p transparência *f*
d Transparenz *f*

* **transparent ice → 518**

2793 transpiration
f transpiration *f*
e transpiración *f*
i traspirazione *f*
p transpiração *f*
d Transpiration *f*

2794 transverse wave
f onde *f* transversale
e onda *f* transversal
i onda *f* trasversale
p onda *f* transversal
d Transversalwelle *f*

2795 travel time
f temps *m* de parcours
e tiempo *m* recorrido
i tempo *m* del percorso
p tempo *m* de percurso
d Laufzeit *f*

2796 triangulation
f triangulation *f*
e triangulación *f*
i triangolazione *f*
p triangulação *f*
d Triangulation *f*

2797 triple collision; three-body collision
f choc *m* triple; collision *f* triple
e colisión *f* tripla
i collisione *f* tripla
p colisão *f* tripla
d Dreierstoß *m*

2798 tritium air monitor
f moniteur *m* atmosphérique de tritium
e monitor *m* atmosférico de tritio
i monitore *m* atmosferico di tritio
p monitor *m* atmosférico de trítio
d Tritiumluftmonitor *m*

2799 tritium surface contamination meter
f contaminamètre *m* surfacique pour tritium
e medidor *m* de contaminación superficial para tritio
i contaminametro *m* di superficie per tritio

p medidor *m* de contaminação da superfície por trítio
d Gerät *n* zur Bestimmung der Tritiumoberflächenkontamination

2800 tropic
f tropique *m*
e trópico *m*
i tropico *m*
p trópico *m*
d Wendekreis *m*

2801 tropical *adj*
f tropical
e tropical
i tropicale
p tropical
d tropisch

2802 tropical agriculture
f agriculture *f* tropicale
e agricultura *f* tropical
i agricoltura *f* tropicale
p agricultura *f* tropical
d Tropenlandwirtschaft *f*

2803 tropical agronomy
f agronomie *f* tropicale
e agronomía *f* tropical
i agronomia *f* tropicale
p agronomia *f* tropical
d Tropenlandwirtschaftslehre *f*

2804 tropical air
f air *m* tropical
e aire *m* tropical
i aria *f* tropicale
p ar *m* tropical
d tropische Luft *f*

2805 tropical and subtropical fruit
f fruits *mpl* tropicaux et subtropicaux
e frutos *mpl* tropicales y meridionales
i frutta *f* tropicale e agrumi *mpl*
p frutos *mpl* tropicais e subtropicais
d Südfrüchte *fpl*

2806 tropical calms
f calmes *mpl* tropicaux
e calmas *fpl* tropicales
i calme *fpl* tropiche
p calmas *fpl* tropicais
d tropische Kalmen *fpl*

2807 tropical climate
f climat *m* tropique
e clima *m* tropical
i clima *m* tropicale
p clima *m* tropical
d tropisches Klima *n*

2808 tropical cyclone; tropical revolving storm; revolving storm
f cyclone *m* tropical
e ciclón *m* tropical
i ciclone *m* tropicale
p ciclone *m* tropical
d tropischer Wirbelsturm *m*

2809 tropical diarrhea
f diarrhée *f* tropicale
e diarrea *f* tropical
i diarrea *f* tropicale
p diarréia *f* tropical
d Tropendiarrhöe *f*

2810 tropical disease
f maladie *f* des pays tropicaux
e enfermedad *f* trópica
i malattia *f* tropica
p enfermidade *f* tropical
d Tropenkrankheit *f*

2811 tropical disturbance
(cyclonic wind system occurring in the tropics but not having the intensity of a tropical cyclone)
f perturbation *f* tropicale
e disturbio *m* atmosférico tropical
i perturbazione *f* tropicale
p distúrbio *m* tropical
d tropische Störung *f*

2812 tropical fever
f fièvre *f* pernicieuse
e fiebre *f* tropical
i febbre *f* tropicale
p febre *f* tropical
d Tropenfieber *n*

2813 tropicalize *v*
f tropicaliser
e tropicalizar
i tropicalizzare
p tropicalizar
d tropikalisieren

2814 tropical malaria
f paludisme *m* tropical
e malaria *f* tropical
i febbre *f* dei tropici
p malária *f* tropical
d Tropenmalaria *f*

2815 tropical maritime air
f air *m* tropical maritime
e aire *m* marítimo tropical
i aria *f* marittima tropicale
p ar *m* marítimo tropical
d tropische feuchte Meeresluft *f*

2816 tropical medicine
f médecine *f* tropicale
e medicina *f* tropical
i medicina *f* tropicale
p medicina *f* tropical
d Tropenmedizin *f*

2817 tropical plant
f plante *f* tropicale
e planta *f* tropical
i pianta *f* tropicale
p planta *f* tropical
d Tropenpflanze *f*

2818 tropical rain forest
f forêt *f* pluviale équatoriale
e selva *f* ecuatorial
i foresta *f* equatoriale
p floresta *f* equatorial
d Regenwald *m*

* **tropical region → 2822**

* **tropical revolving storm → 2808**

2819 tropical tide
f marée *f* des tropiques
e marea *f* de los trópicos
i marea *f* dei tropici
p maré *f* dos trópicos
d Tropentide *f*

2820 tropical wind
f vent *m* des tropiques
e viento *m* trópico
i vento *m* tropico
p vento *m* tropical
d tropischer Wind *m*

2821 tropical year
f année *f* tropicale
e año *m* tropical
i anno *m* tropicale
p ano *m* tropical
d tropisches Jahr *n*

2822 tropical zone; tropical region
f région *f* tropicale
e zona *f* tropical; región *f* tropical
i zona *f* tropicale; regione *f* tropicale
p zona *f* tropical; região *f* tropical
d Tropenzone *f*; tropische Gegend *f*

2823 tropic of Cancer
f tropique *m* du Cancer
e trópico *m* de Cáncer
i tropico *m* del Cancro
p trópico *m* de Câncer
d Wendekreis *m* des Krebses

2824 tropic of Capricorn
f tropique *m* du Capricorne
e trópico *m* de Capricornio
i tropico *m* del Capricornio
p trópico *m* de Capricórnio
d Wendekreis *m* des Steinbocks

2825 tropopause
(the atmospheric region between troposphere and stratosphere)
f tropopause *f*
e tropopausa *f*
i tropopausa *f*
p tropopausa *f*
d Tropopause *f*

2826 troposphere
(the atmospheric region nearest to earth)
f troposphére *f*
e troposfera *f*
i troposfera *f*
p troposfera *f*
d Troposphäre *f*

2827 tropospheric air movement
f mouvement *m* d'air troposphérique
e movimiento *m* de aire troposférica
i movimento *m* d'aria troposferica
p movimento *m* de ar troposférico
d troposphärische Luftbewegung *f*

2828 tropospheric fall-out
f retombée *f* troposphérique
e depósito *m* troposférico
i ricaduta *f* troposferica
p precipitação *f* troposférica radioativa
d Troposphärenausfall *m*

2829 trough
f creux *m* barométrique
e vaguada *f*
i avvallamento *m*
p depressão *f* oceânica
d Tiefdruckrinne *f*

2830 true height; real height
f hauteur *f* vraie; hauteur *f* réelle
e altitud *f* verdadera
i altezza *f* vera; altezza *f* reale
p altura *f* verdadeira
d wahre Höhe *f*

2831 true midnight
f minuit *m* vrai
e medianoche *f* verdadera
i mezzanotte *f* vera
p meia-noite *f* verdadeira
d wahre Mitternacht *f*

2832 true north
f nord *m* vrai
e norte *m* verdadero
i nord *m* vero
p norte *m* verdadeiro
d rechtweisender Nord *m*

2833 true solar day
f jour *m* solaire vrai
e día *m* solar verdadero
i giorno *m* solare vero
p dia *m* solar verdadeiro
d wahrer Sonnentag *m*

*** true sun → 186**

2834 true time
f temps *m* vrai
e tiempo *m* verdadero
i tempo *m* vero
p tempo *m* verdadeiro
d wahre Zeit *f*

2835 truncation error
f erreur *f* de troncature
e error *m* de truncamiento
i errore *m* di troncamento
p erro *m* de truncamento
d Rundungsfehler *m*

2836 tundra
(vegetation covering large areas in arctic and subarctic regions)
f toundra *f*
e tundra *f*
i tundra *f*
p tundra *f*
d Tundra *f*

2837 tundra climate
f climat *m* des toundras
e clima *m* de la tundra
i clima *m* da tundra
p clima *m* da tundra
d Tundrenklima *n*

2838 turbidity
f trouble *m* atmosphérique
e turbiedad *f*
i torbidità *f*
p turbidez *f*
d Trübung *f*

2839 turbosphere
f turbosphère *f*
e turbosfera *f*
i turbosfera *f*
p turbosfera *f*
d Turbosphäre *f*

*** turbulence → 1292**

2840 turbulence cloud
f nuage *m* de turbulence
e nube *f* turbulenta
i nube *f* dovuta a turbolenza
p nuvem *f* de turbulência
d Wirbelwolke *f*

*** turbulent air → 1292**

2841 turbulent boundary layer
f couche *f* limite turbulente
e capa *f* límite turbulenta
i strato *m* limite turbolento
p camada *f* limite turbulenta
d turbulente Grenzschicht *f*

2842 turbulent flow
f courant *m* turbulent
e corriente *f* turbulenta
i corrente *f* turbolenta
p corrente *f* turbulenta
d turbulente Strömung *f*

*** turn of the tide → 495**

2843 twilight; dusk
f crépuscule *m*
e crepúsculo *m*
i crepuscolo *m*
p crepúsculo *m*
d Zweilicht *n*

2844 twilight arc
f arc *m* crépusculaire
e arco *m* crepuscular
i arco *m* crepuscolare
p arco *m* crepuscular
d Dämmerungsbogen *m*

2845 twilight phenomenon
f phénomène *m* crépusculaire
e fenómeno *m* del crepúsculo
i fenomeno *m* crepuscolare
p fenômeno *m* do crepúsculo
d Dämmerungserscheinung *f*

2846 twilight sky
f ciel *m* crépusculaire
e cielo *m* crepuscular
i cielo *m* crepuscolare
p céu *m* crepuscular
d Dämmerungshimmel *m*

2847 twister
f tornado *m* violent
e huracán *m* de corta duración
i tromba *f* d'aria
p pé *m* de vento
d Windhose *f*

2848 type of soil
f sorte *f* du sol
e tipo *m* de suelo
i tipo *m* di terreno
p tipo *m* do solo
d Bodenart *f*

2849 types of precipitation
f sortes *fpl* de précipitation
e tipos *mpl* de precipitaciones
i tipi *mpl* di precipitazioni
p tipos *mpl* de precipitações
d Niederschlagsformen *fpl*

2850 typhoon
f typhon *m*
e tifón *m*
i tifone *m*
p tufão *m*
d Taifun *m*

U

2851 ugly weather
f temps *m* menaçant
e tiempo *m* amenazante
i tempo *m* minacciante
p tempo *m* feio
d drohendes Wetter *n*

2852 ultraviolet radiation
f rayonnement *m* ultraviolet
e radiación *f* ultravioleta
i radiazione *f* ultravioletta
p radiação *f* ultravioleta
d ultraviolette Strahlung *f*

2853 ultraviolet rays
f rayons *mpl* ultraviolets
e rayos *mpl* ultravioleta
i raggi *mpl* ultravioletti
p raios *mpl* ultravioletas
d ultraviolette Strahlen *mpl*

2854 Umkehr effect
f effet *m* Umkehr
e efecto *m* de Umkehr
i effetto *m* d'inversione; effetto *m* d'Umkehr
p efeito *m* de Umkehr
d Umkehreffekt *m*

2855 uncinus
f uncinus *m*
e uncinus *m*
i uncino *m*
p uncinus *m*
d Uncinus *m*

2856 uncloudedness; clearness
f clarté *f*
e claridad *f*
i chiarezza *f*
p claridade *f*
d Klarheit *f*

2857 undercurrent
f courant *m* sous-marin
e corriente *f* submarina
i sottocorrente *f*
p subcorrente *f*
d Unterströmung *f*

2858 underset; undertow
f contre-courant *m* sous-marin
e contracorriente *f* submarina
i controcorrente *f* sottomarina
p contracorrente *f* submarina
d Unterwasserstrom *m*

2859 under steam
f sous vapeur
e bajo vapor
i sotto vapore
p baixo vapor
d unter Dampf

*** undertow → 2858**

2860 unfavourable weather
f temps *m* défavorable
e tiempo *m* desfavorable
i tempo *m* sfavorevole
p tempo *m* desfavorável
d ungünstiges Wetter *n*

2861 unit
f unité *f*
e unidad *f*
i unità *f*
p unidade *f*
d Einheit *f*

*** universal time → 1269**

2862 unproductive areas
f terres *fpl* improductives
e tierras *fpl* improductivas
i terre *fpl* improduttive
p terras *fpl* improdutivas
d ertragsunfähige Flächen *fpl*

2863 unsaturated air
f air *m* non saturé
e aire *m* no saturado
i aria *f* non satura
p ar *m* não saturado
d ungesättigte Luft *f*

2864 unsettled weather
f temps *m* incertain
e tiempo *m* incierto
i tempo *m* incerto
p tempo *m* incerto
d unbeständiges Wetter *n*

2865 unstable layer
f couche *f* instable
e capa *f* inestable
i strato *m* instabile
p camada *f* instável
d labile Schicht *f*

*** up-current → 133**

2866 upper atmosphere; high atmosphere
f atmosphère *f* supérieure; haute atmosphère *f*
e atmósfera *f* alta
i atmosfera *f* superiore
p atmosfera *f* superior
d obere Atmosphäre *f*; Hochatmosphäre *f*

2867 upper limit of ozone layer
f limite *f* supérieure de la couche d'ozone
e límite *m* superior de la capa de ozono
i limite *m* superiore del strato di ozono
p limite *m* superior da camada de ozônio
d Obergrenze *f* der Ozonschicht

2868 upper tangential arc
f arc *m* tangent supérieur
e arco *m* tangente superior
i arco *m* tangente superiore
p arco *m* tangente superior
d oberer Berührungsbogen *m*

2869 upper wind
f vent *m* en altitude
e viento *m* en altura
i vento *m* in quota
p vento *m* em altitude
d Höhenwind *m*

2870 upper wind report
f message *m* aérologique
e mensaje *f* de sondeo del viento
i bollettino *m* aerologico
p boletim *m* aerológico
d Höhenwindmessung *f*

* **upslope fog → 1367**

V

2871 valley
f vallée *f*
e valle *m*
i valle *m*
p vale *m*
d Tal *n*

2872 valley breeze
f brise *f* de vallée
e brisa *f* de valle
i brezza *f* di valle
p brisa *f* do vale
d Talwind *m*

2873 valley glacier
f glacier *m* de vallée
e glaciar *m* de valle
i ghiaccio *m* di vallone
p glaciar *m* de vale
d Talgletscher *m*

2874 Van Allen layer
f zone *f* ionosphérique Van Allen
e zona *f* ionosférica Van Allen
i zona *f* ionosferica Van Allen
p zona *f* ionosférica Van Allen
d Van Allengürtel *m*

2875 vaporization
f vaporisation *f*
e vaporación *f*
i vaporizzazione *f*
p vaporização *f*
d Verdampfung *f*

2876 vapour pressure; vapour tension
f pression *f* de vapeur
e presión *f* de vapor
i pressione *f* di vapore
p pressão *f* de vapor
d Verdunstungsdruck *m*

2877 vapour pressure thermometer
f thermomètre *m* à pression de vapeur
e termómetro *m* de presión del vapor
i termometro *m* a pressione di vapore
p termômetro *m* de pressão de vapor
d Dampfdruckthermometer *n*

*** vapour tension → 2876**

2878 variable *adj*
f variable
e variable
i variabile
p variável
d veränderlich

2879 variable weather
f temps *m* variable
e tiempo *m* variable
i tempo *m* variabile
p tempo *m* variável
d veränderliches Wetter *n*

2880 variable winds
f vents *mpl* variables
e vientos *mpl* variables
i venti *mpl* variabili
p ventos *mpl* variáveis
d wechselnde Winde *mpl*

2881 variance
f variance *f*
e varianza *f*
i varianza *f*
p variância *f*
d Veränderung *f*

2882 variation of temperature; temperature range
f variation *f* de température
e variación *f* de temperatura
i variazione *f* di temperatura
p variação *f* de temperatura
d Temperaturschwankung *f*

2883 variation of terrestrial magnetism
f variation *f* du magnétisme terrestre
e variación *f* del magnetismo terrestre
i variazione *f* del magnetismo terrestre
p variação *f* do magnetismo terrestre
d erdmagnetische Variation *f*

2884 variometer
f variomètre *m*
e variómetro *m*
i variometro *m*
p variômetro *m*
d Variometer *n*

2885 veer *v*
f changer de direction vers la droite
e arribar
i cambiare da sinistra a dritta
p mudar de direção para a direita
d mit der Sonne drehen

2886 veer *v* **aft**
f adonner
e alagarse

i ridondare
p alargar-se
d raumen

2887 veering; backing
(clockwise change in direction of wind)
f rotation *f* vers la droite
e cambio *m* dextrógiro
i vento *m* girante verso destra
p mudança *f* dextrógira
d Rechtsdrehen *n*

2888 veering of wind
f virement *m* du vent
e rolada *f* del viento
i vento *m* girante
p vento *m* dextrógiro
d Winddrehung *f*

2889 Vegard-Kaplan bands
f bandes *fpl* de Vegard-Kaplan
e bandas *fpl* de Vegard-Kaplan
i bande *fpl* di Vegard-Kaplan
p bandas *fpl* de Vegard-Kaplan
d Vegard-Kaplan-Banden *fpl*

2890 vegetation; plant cover
f végétation *f*
e vegetación *f*; cubierta *f* vegetal
i vegetazione *f*; copertura *f* vegetale
p vegetação *f*
d Vegetation *f*; Pflanzendecke *f*

2891 vegetation cone
f cône *m* de végétation
e cono *m* vegetativo
i cono *m* vegetativo
p cone *m* vegetativo
d Vegetationskegel *m*

* **vegetation period → 1285**

2892 vegetative point
f point *m* de végétation
e punto *m* vegetativo
i punto *m* vegetativo
p ponto *m* de vegetação
d Vegetationspunkt *m*

2893 vegetative rest
f repos *m* de la végétation
e vegetación *f* latente
i riposo *m* vegetativo
p vegetação *f* latente
d Vegetationsruhe *f*

2894 veil of cloud; scarf cloud
f voile *m* de nuages
e cirro *m* de plumas
i velo *m* di nubi
p véu *m* de nuvens
d Wolkenschleier *m*

2895 velocity
f vitesse *f*
e velocidad *f*
i velocità *f*
p velocidade *f*
d Geschwindigkeit *f*

2896 vernal *adj*
f vernal
e vernal; primaveril
i primaverile
p vernal
d Frühlings...

2897 vernal equinox
f équinoxe *m* de printemps
e equinoccio *m* vernal
i equinozio *m* di primavera
p equinócio *m* de primavera
d Frühlingsnachtgleiche *f*

2898 vernalization
f jarovisation *f*
e determinación *f* germinal
i jarovizzazione *f*
p vernalização *f*
d Jarowisation *f*

2899 vernal point
f point *m* vernal
e punto *m* vernal
i punto *m* primaverile
p ponto *m* vernal
d Frühlingspunkt *m*

2900 vernier
f vernier *m*
e nonio *m*
i verniero *m*
p nônio *m*
d Nonius *m*

2901 vertical current of air
f courant *m* d'air vertical
e corriente *f* de aire vertical
i corrente *f* d'aria verticale
p corrente *f* de ar vertical
d vertikaler Luftstrom *m*

2902 vertical distribution
f distribution *f* verticale
e distribución *f* vertical
i distribuzione *f* verticale
p distribuição *f* vertical
d Vertikalverteilung *f*

2903 vertical intensity
f intensité *f* verticale
e intensidad *f* vertical
i intensità *f* verticale
p intensidade *f* vertical
d Vertikalintensität *f*

2904 very long waves
f ondes *fpl* très longues
e ondas *fpl* muy largas
i onde *fpl* lunghissime
p ondas *fpl* muito longas
d Längstwellen *fpl*

2905 very low frequency
f très basse fréquence *f*
e hipofrecuencia *f*
i frequenza *f* molto bassa
p frequência *f* muito baixa
d sehr niedrige Frequenz *f*

2906 vinery
f serre *f* à vigne
e estufa *f* para vides
i serra *f* da vite
p estufa *f* para videiras
d Traubenhaus *n*

2907 virgin forest; primeval forest
f forêt *f* vierge
e selva *f* virgen
i foresta *f* vergine
p floresta *f* virgem
d Urwald *m*

* **virtual height → 181**

2908 viscosity
f viscosité *f*
e viscosidad *f*
i viscosità *f*
p viscosidade *f*
d Zähigkeit *f*

2909 visibility
f visibilité *f*
e visibilidad *f*
i visibilità *f*
p visibilidade *f*
d Sicht *f*

2910 visual meteorological conditions
f conditions *fpl* météorologiques visuelles
e condiciones *fpl* meteorológicas visuales
i condizioni *fpl* meteorologiche visibili
p condições *fpl* meteorológicas a olho nu
d sichtbare meteorologische Bedingungen *fpl*

2911 visual range
f portée *f* visuelle
e alcance *m* visual
i portata *f* visuale
p alcance *m* visual
d visuelle Reichweite *f*

2912 volcanic activity
f activité *f* volcanique
e actividad *f* volcánica
i attività *f* vulcanica
p atividade *f* vulcânica
d vulkanische Tätigkeit *f*

2913 volcanic ash
f cendre *f* volcanique
e ceniza *f* volcánica
i cenere *f* vulcanica
p cinza *f* vulcânica
d vulkanische Asche *f*

2914 volcanic bomb
f bouchon *m* volcanique
e bomba *f* volcánica
i bomba *f* vulcanica
p bomba *f* vulcânica
d Vulkanbombe *f*

2915 volcanic dust
f poussière *f* volcanique
e polvo *m* volcánico
i polvere *f* vulcanica
p poeira *f* vulcânica
d vulkanischer Staub *m*

2916 volcanic eruption
f explosion *f* volcanique
e explosión *f* volcánica
i esplosione *f* vulcanica
p explosão *f* vulcânica
d Ausbruchserscheinung *f*

2917 volcanic soil
f terre *f* volcanique
e tierra *f* volcánica
i terreno *m* vulcanico
p solo *m* vulcânico
d vulkanischer Boden *m*

2918 volcanic tuffa
f tuf *m* volcanique
e tufa *f* volcánica
i tufo *m* vulcanico
p tufo *m* vulcânico
d vulkanischer Tuffstein *m*

2919 volcano
f volcan *m*
e volcán *m*

i vulcano *m*
p vulcão *m*
d Vulkan *m*

* **vortex → 912**

2920 vortical motion; eddy motion
f mouvement *m* tourbillonnaire
e movimiento *m* de remolino
i moto *m* turbolento
p movimento *m* turbilhonar; movimento *m* vorticoso
d Wirbelbewegung *f*

2921 vorticity
f tourbillon *m*
e vorticidad *f*
i vorticità *f*
p vorticidade *f*
d Wirbelung *f*

2922 wandering dunes
f dunes *fpl* mobiles
e dunas *fpl* móviles
i dune *fpl* mobili
p dunas *fpl* móveis
d Wanderdünen *fpl*

2923 waning moon; decrescent moon
f lune *f* décroissante
e luna *f* menguante
i luna *f* calante
p lua *f* minguante
d abnehmender Mond *m*

2924 warm air mass
f masse *f* d'air chaud
e masa *f* de aire caliente
i massa *f* d'aria calda
p massa *f* de ar quente
d Warmluftmasse *f*

2925 warm current
f courant *m* chaud
e corriente *f* caliente
i corrente *f* calda
p corrente *f* quente
d warmer Strom *m*

2926 warm front
f front *m* chaud
e frente *m* calido
i fronte *m* caldo
p frente *f* quente
d Warmluftfront *f*

2927 warming of the air
f chauffage *m* de l'air
e calentamiento *m* del aire
i riscaldamento *m* dell'aria
p aquecimento *m* do ar
d Erwärmung *f* der Luft

2928 warm ocean current
f courant *m* marin chaud
e corriente *f* marina cálida
i corrente *f* oceanica calda
p corrente *f* marítima quente
d warme Meeresströmung *f*

2929 warm sector
f secteur *m* chaud
e sector *m* calido
i settore *m* caldo
p setor *m* quente
d Warmsektor *m*

2930 warm weather
f temps *m* chaud
e tiempo *m* caliente
i tempo *m* caldo
p tempo *m* quente
d warmes Wetter *n*

*** waste land → 1500**

2931 water
f eau *f*
e agua *f*
i acqua *f*
p água *f*
d Wasser *n*

*** water accumulation → 2951**

2932 water balance
f balance *f* hydrique
e balance *m* hídrico
i bilancio *m* idrico
p equilíbrio *m* de água
d Wasserbilanz *f*

2933 water barometer
f baromètre *m* à eau
e barómetro *m* de agua
i barometro *m* d'acqua
p barômetro *m* de água
d Wasserbarometer *n*

2934 water consumption
f consommation *f* d'eau
e consumo *m* de agua
i consumo *m* d'acqua
p consumo *m* de água
d Wasserverbrauch *m*

2935 water content
f teneur *m* en eau
e contenido *m* en agua
i contenuto *m* idrico
p teor *m* de água
d Wassergehalt *m*

2936 watercourse
f cours *m* d'eau
e corriente *f* de agua
i corso *m* d'acqua
p curso *m* de água
d Wasserlauf *m*

*** water deficiency area → 215**

2937 water development and use; water utilization
f régime *m* des eaux
e régimen *m* de aguas
i regime *m* idrico
p regime *m* de águas
d Wasserwirtschaft *f*

2938 water economy
f économie *f* hydraulique
e economía *f* de agua
i economia *f* idrica
p economia *f* hidráulica
d Wasserhaushalt *m*

2939 water erosion
f érosion *f* par l'eau
e erosión *f* por el agua
i erosione *f* da acqua
p erosão *f* provocada pela água
d Wassererosion *f*

* **water free of ice → 1972**

2940 water gauge
f échelle *f* d'étiage
e indicador *m* del nivel de la agua
i indicatore *m* di livello dell'acqua
p indicador *m* do nível de água
d Wasserstandszeiger *m*

2941 waterholding capacity
f capacité *f* de rétention de l'eau
e capacidad *f* de fijación de agua
i capacità *f* di ritenzione idrica
p capacidade *f* de retenção de água
d Wasserbindungsvermögen *n*

* **watering → 1540**

2942 water inlet
f adduction *f* d'eau
e acometida *f* de agua
i adduzione *f* d'acqua
p entrada *f* de água
d Wasserzuleitung *f*

2943 waterlogging
f saturation *f* du sol par l'eau
e saturación *f* del suelo por el agua
i saturazione *f* acquosa del terreno
p saturação *f* do solo pela água
d Vernässung *f*

2944 water movement
f mouvement *m* de l'eau
e movimiento *m* del agua
i moto *m* dell'acqua; circolazione *f* dell'acqua
p movimento *m* de água
d Wasserbewegung *f*

2945 water protection
f protection *f* des eaux
e protección *f* acuosa
i protezione *f* delle acque
p proteção *f* das águas
d Gewässerschutz *m*

2946 water raising
f élévation *f* d'eau
e elevación *f* de aguas
i sollevamento *m* dell'acqua
p elevação *f* das águas
d Wasserförderung *f*

2947 water right
f droit *m* en matière d'eaux
e derecho *m* de aguas
i diritto *m* in materia di acque
p direito *m* de águas
d Wasserrecht *n*

2948 watershed
f ligne *f* de partage des eaux
e divisoria *f* de las aguas
i divisione *f* dell'acqua
p linha *f* divisória de águas
d Wasserscheide *f*

2949 water sky
f ciel *m* de clairière
e cielo *m* de agua
i cielo *m* d'acqua
p céu *m* d'água
d Wasserhimmel *m*

2950 waterspout
f trombe *f* marine
e tromba *f* marina
i tromba *f* marina; tromba *f* d'acqua
p tromba *f* de água
d Wasserhose *f*

2951 water storage; water accumulation
f emmagasinage *m* des eaux
e acumulación *f* de agua
i raccolta *f* delle acque
p reservatório *m* de água
d Wasser(auf)speicherung *f*

2952 water table
f niveau *m* de l'eau
e nivel *m* del agua
i superficie *f* freatica; livello *m* dell'acqua; falda *f*
p camada *f* freática
d Wasserspiegel *m*

2953 watertight *adj*
f étanche à l'eau
e estanco al agua
i stagno all'acqua
p estanque
d wasserdicht

* **water utilization → 2937**

2954 water vapour; aqueous vapour
f vapeur *f* d'eau
e vapor *m* acuoso
i vapore *m* d'acqua
p vapor *m* d'água
d Wasserdampf *m*

* **water-vapour pressure → 2131**

2955 wave
f vague *f*
e ola *f*
i onda *f*
p onda *f*
d Welle *f*

2956 wave cyclone
f cyclone *m* à ondulations
e ciclón *m* de onda
i ciclone *m* d'onda
p ciclone *m* com ondulações
d Wellenzyklon *m*

2957 wave height; height of wave
f hauteur *f* de vague
e altura *f* de ola
i altezza *f* d'onda
p altura *f* de onda
d Wellenhöhe *f*

2958 wave motion
f mouvement *m* en vague
e movimiento *m* ondulatorio
i moto *m* ondoso
p movimento *m* ondulatório
d Wellenbewegung *f*

2959 wave period
f période *f* de vague
e período *m* de ola
i periodo *m* d'onda
p período *m* de onda
d Wellenperiode *f*

2960 weather *v*
f se dégrader
e descomponerse *(por las inclemencias atmosféricas)*
i degradarsi *(alle intemperie)*
p desintegrar-se *(pela ação do tempo)*
d verwittern

2961 weather analysis
f analyse *f* du temps
e análisis *f* del tiempo
i analisi *f* del tempo
p análise *f* do tempo
d Wetteranalyse *f*

2962 weather-beaten *adj*
f battu par la mer
e acosado por el mal tiempo
i battuto dalla tempesta
p batido pelo tempo
d verwittert

2963 weather-bound *adj*
f arrêté par le gros temps
e detenido por mal tiempo
i fermato del cattivo tempo
p detido pelo mau tempo
d durch Unwetter zurückgehalten

* **weather change → 496**

* **weather chart → 1804**

2964 weather coast
f côte *f* au vent
e costa *f* de barlovento
i costa *f* sopravvento
p costa *f* de barlavento
d Wetterküste *f*

2965 weather conditions; meteorological conditions
f conditions *fpl* météorologiques
e condiciones *fpl* meteorológicas
i condizioni *fpl* meteorologiche
p condições *fpl* meteorológicas
d Wetterverhältnisse *npl*

2966 weather damage
f dommage *m* causé par le mauvais temps
e daño *m* causado por mal tiempo
i danno *m* causato da intemperie
p dano *m* causado pelo mal tempo
d Witterungsschaden *m*

2967 weather forecast
f prévision *f* du temps
e predicción *f* del tiempo
i previsione *f* del tempo
p previsão *f* do tempo
d Wettervorhersage *f*

* **weather, in any sort of ~ → 1488**

2968 weathering
f altération *f* superficielle par les agents atmosphériques
e desgaste *m* por la intemperie
i alterazione *f* superficiale
p desgaste *m* por intempérie
d Verwitterung *f*

2969 weather maxim
f proverbe *m* météorologique
e refrán *m* meteorológico
i aforismo *m* sul tempo
p máximas *fpl*
d Wetterregel *n*; Wetterspruch *m*

2970 weather message
f avertissement *m* météorologique
e información *f* meteorológica
i informazione *f* meteorologica
p informação *f* meteorológica
d Wettermeldung *f*

2971 weather *v* **out**
f étaler
e aguantar
i sostenere
p resistir
d aushalten

2972 weather *v* **out a gale**
f étaler un coup de vent
e aguantar un viento violento
i sostenere un colpo di vento
p resistir à borrasca
d einen Sturm abreiten

2973 weatherproof *adj*
f à l'épreuve d'intempérie
e resistente a la intemperie
i inattaccabile sotto le intemperie
p à prova de intempéries do tempo
d wetterfest

2974 weather radar
f radar *m* météorologique
e radar *m* meteorológico
i radar *m* meteorologico
p radar *m* de tempo
d Niederschlagsradar *n*

2975 weather report; meteorological report
f bulletin *m* météorologique
e boletín *m* meteorológico
i bollettino *m* meteorologico
p boletim *m* meteorológico
d Wetterbericht *m*

2976 weather satellite
f satellite *m* météorologique
e satélite *m* meteorológico
i satellite *m* meteorologico
p satélite *m* meteorológico
d Wettersatellit *m*

2977 weather signs
f signes *mpl* du temps
e indicios *mpl* de tiempo
i indizi *mpl* di tempo
p indícios *mpl* de tempo
d Wetteranzeichen *npl*

2978 weather station; meteorological observatory
f station *f* météorologique; observatoire *m* météorologique
e estación *f* meteorológica; observatorio *m* meteorológico
i stazione *f* meteorologica; osservatorio *m* meteorologico
p estação *f* meteorológica; observatório *m* meteorológico
d Wetterwarte *f*; meteorologische Station *f*

2979 weather window
f fenêtre *f* météorologique
e ventana *f* meteorológica
i finestra *f* meteorologica
p janela *f* meteorológica
d Wetterfenster *n*

2980 wedge of high pressure
f crête *f* de haute pression
e cuña *f* de alta presión
i cuneo *m* di alta pressione
p cunha *f* de alta pressão
d Hochdruckkeil *m*

2981 weighted average
f moyenne *f* pondérée
e media *f* ponderada
i media *f* ponderata
p média *f* ponderada
d gewogener Durchschnitt *m*

2982 weight of the atmosphere
f poids *m* de l'atmosphère
e peso *m* de la atmósfera
i peso *m* dell'atmosfera
p peso *m* da atmosfera
d Gewicht *m* der Atmosphäre

2983 well ventilated *adj*
f bien aéré
e airoso
i arioso
p airoso
d gut belüftet

2984 west
f ouest *m*
e oeste *m*
i ovest *m*
p oeste *m*
d Westen *m*

2985 West Australian Current
f Courant *m* australien occidental
e Corriente *f* australiana occidental
i Corrente *f* australiana occidentale
p Corrente *f* australiana ocidental
d Westaustralstrom *m*

2986 westerlies
f vents *mpl* d'ouest
e vientos *mpl* del oeste
i correnti *fpl* occidentali
p ventos *mpl* de oeste
d westliche Winde *mpl*

2987 westerly hour angle
f angle *m* horaire occidental
e ángulo *m* horario occidental
i angolo *m* orario occidentale
p ângulo *m* horário ocidental
d westlicher Stundenwinkel *m*

*** westerly wind → 2991**

2988 western amplitude
f amplitude *f* occidentale
e amplitud *f* occidental
i amplitudine *f* al tramonto
p amplitude *f* ocidental
d Abendweite *f*

2989 western declination
f déclinaison *f* occidentale
e declinación *f* occidental
i declinazione *f* ovest
p declinação *f* ocidental
d westliche Deklination *f*

2990 West European time
f heure *f* de l'Europe occidentale
e hora *f* de la Europa occidental
i ora *f* dell'Europa occidentale
p hora *f* da Europa ocidental
d westeuropäische Zeit *f*

2991 west wind; westerly wind
f vent *m* d'ouest
e viento *m* del oeste
i vento *m* di ovest
p vento *m* de oeste
d Westwind *m*

2992 west wind drift; antarctic circumpolar current
f dérive *f* du vent d'ouest
e corriente *f* de viento del oeste
i corrente *f* circumpolare antartica
p corrente *f* de vento de oeste
d Westwinddrift *f*

2993 wet-bulb thermometer
f thermomètre *m* mouillé
e termómetro *m* de bulbo húmedo
i termometro *m* a bulbo bagnato
p termômetro *m* de bulbo úmido
d Naßthermometer *n*

2994 wet fog
f brouillard *m* humide
e niebla *f* húmeda
i nebbia *f* umida
p nevoeiro *m* úmido
d nässender Nebel *m*

2995 wetness of steam
f humidité *f* de la vapeur
e humedad *f* del vapor
i umidità *f* del vapore
p umidade *f* do vapor
d Dampfnässe *f*

2996 wet snow; damp snow
f neige *f* humide
e nieve *f* húmeda
i neve *f* bagnata; neve *f* umida
p neve *f* úmida
d Naßschnee *m*; Feuchtschnee *m*

*** wet soil → 1235**

2997 wettability
f propriété *f* mouillante
e humectabilidad *f*
i bagnabilità *f*
p umectabilidade *f*
d Netzfähigkeit *f*

2998 wetted surface
f surface *f* mouillée
e superficie *f* bañada
i superficie *f* bagnata
p superfície *f* molhada
d benetzte Oberfläche *f*

2999 wetting agent
f agent *m* mouillant
e agente *m* humectante
i agente *m* umettante
p agente *m* umedecedor
d Benetzungsmittel *n*

3000 wheel barometer
f baromètre *m* à cadran
e barómetro *m* indicador
i barometro *m* a quadrante
p barômetro *m* de mostrador
d Zeigerbarometer *n*

* **whirlpool → 912**

3001 whirlwind; tornado
f tourbillon *m* de vent; tornado *m*
e torbellino *m* de viento; tornado *m*
i vento *m* vorticoso; tornado *m*
p redemoinho *m* de vento
d Wirbelwind *m*; Tornado *m*

3002 whistlers; whistling atmospherics
f siffleurs *mpl*; sifflements *mpl*
e silbadores *mpl*
i fischiatori *mpl*
p assobios *mpl*
d atmosphärische Pfeifstörungen *fpl*

* **whistling atmospherics → 3002**

3003 white cap
f capuchon *m* blanc
e cabretilla *f*
i onda *f* con cresta bianca
p carneiro *m*
d Schaumkämme *m*

* **white ice → 1047**

3004 white rainbow; fogbow; fogdog
f arc-en-ciel *m* blanc
e arco *m* iris blanco
i arcobaleno *m* bianco
p arco-íris *m* branco
d weißer Regenbogen *m*

3005 white squall
f grain *m* blanc
e chubasco *m* blanco
i groppo *m* chiaro
p tormenta *f* branca
d weiße Bö *f*

3006 whole gale
f tempête *f*
e temporal *m*
i temporale *m* fortissimo
p vento *m* muito duro
d starker Sturm *m*

3007 willy-willy
(tropical cyclone in Australia)
f willy-willy *m*
e willy-willy *m*
i willy-willy *m*
p willy-willy *m*
d Wirbelsturm *m*

3008 wilting point
f point *m* de flétrissement
e punto *m* permanente de apergaminamiento
i punto *m* di avvizzimento
p ponto *m* de emurchecimento
d Welkepunkt *m*

3009 wind
f vent *m*
e viento *m*
i vento *m*
p vento *m*
d Wind *m*

* **wind across → 676**

3010 wind action
f action *f* éolienne
e acción *f* eólica
i azione *f* eolica
p ação *f* eólica
d Windwirkung *f*

3011 windblown sand
f sable *m* éolien
e arena *f* de duna
i sabbia *f* eolica
p areia *f* arrastada pelo vento
d Flugsand *m*

3012 windbound *adj*
f arrêté par le vent
e detenido por viento contrario
i fermato per il vento
p retido pelo vento
d von konträren Winden aufgehalten

* **windbreak → 3034**

3013 windburn
f brûlure *f* de vent
e quemadura *f* del viento
i danno *m* causato da venti caldi e secchi
p queimadura *f* de vento
d durch kalten Wind geschädigte Hautstelle *f*

3014 wind chart
f carte *f* des vents
e carta *f* de vientos
i carta *f* dei venti
p carta *f* dos ventos
d Windkarte *f*

3015 wind cloud
f nuage *m* de vent
e nube *f* de viento
i nuvola *f* di vento
p nuvem *f* de vento
d Windwolke *f*

* **wind cone → 3022**

3016 wind damage
f dégât *m* causé par le vent
e daño *m* causado por el viento
i dano *m* causato da vento
p dano *m* causado pelo vento
d Windschaden *m*

3017 wind direction; wind's eye
f direction *f* du vent
e dirección *f* del viento
i direzione *f* del vento
p direção *f* do vento
d Windrichtung *f*

3018 wind-direction indicator
f indicateur *m* de direction du vent
e indicador *m* de dirección de viento
i indicatore *m* di direzzione del vento
p indicador *m* de direção do vento
d Windrichtungsanzeiger *m*

3019 wind erosion
f érosion *f* éolienne
e erosión *f* eólica
i erosione *f* eolica
p erosão *f* eólica
d Winderosion *f*

3020 wind force
f force *f* de vent
e fuerza *f* del viento
i forza *f* del vento
p força *f* do vento
d Windstärke *f*

3021 wind force 8
f vent *m* force 8
e viento *m* fuerza 8
i vento *m* forza 8
p vento *m* força 8
d Windstärke *f* 8

* **wind force ..., with a ~ → 3077**

3022 wind hose; wind cone
f manche *f* à vent
e manga *f* de viento
i manica *f* da vento
p manga *f* de vento
d Windsack *m*

* **wind, into the ~ → 1522**

3023 wind is changing
f le vent tourne
e el viento salta
i il vento cambia
p o vento está mudando
d der Wind springt um

3024 wind lulls down
f le vent tombe
e el viento cae
i il vento cade
p o vento amaina
d der Wind flaut ab

* **wind measurement → 145**

3025 windmill
f moulin *m* a vent
e molino *m* de viento
i mulino *m* a vento
p moinho *m* de vento
d Windmühle *f*

3026 windmill anemometer
f anémomètre *m* à moulinet
e anemómetro *m* de molinete
i anemometro *m* a ventola
p anemômetro *m* de molinete
d Windmühleanemometer *n*

3027 wind motion
f mouvement *m* du vent
e movimiento *m* del viento
i moto *m* del vento
p movimento *m* do vento
d Windbewegung *f*

3028 wind pollination
f pollinisation *f* par le vent
e polinización *f* por viento
i impollinazione *f* anemofila
p polinização *f* pelo vento
d Windbestäubung *f*

3029 wind pressure
f pression *f* du vent
e presión *f* del viento
i pressione *f* del vento
p pressão *f* do vento
d Winddruck *m*

3030 wind ripple
f ride *f* de vent
e ripple *m* eólico
i increspatura *f* eolica
p ondulação *f* pelo vento
d Windrippel *m*

* **wind rose → 595**

3031 wind sediment; eolian sediment
f dépôt *m* éolien
e depósito *m* eólico
i deposito *m* eolico
p depósito *m* eólico
d Windablagerung *f*

* **wind's eye → 3017**

3032 wind shear
f cisaillement *m* du vent
e cizalladura *f* del viento
i gradiente *m* del vento
p cisalhamento *m* do vento
d Windscherung *f*

3033 windshed
f ligne *f* de séparation des vents
e línea *f* divisoria de vientos
i spartivento *m*
p linha *f* divisória dos ventos
d Einzugsgebiet *n*

3034 wind shelter; windbreak; shelter belt
(natural protection against wind and erosion formed by trees and shrubs)
f abri *m* contre le vent; brise-vent *m*
e abrigo *m* contra el viento; cortavientos *m*
i protezione *f* contro il vento; impianto *m* frangivento
p abrigo *m* contra o vento; pára-vento *m*
d Windschutz *m*

3035 wind speed; wind velocity
f vitesse *f* du vent
e velocidad *f* del viento
i velocità *f* del vento
p velocidade *f* do vento
d Windgeschwindigkeit *f*

3036 wind transport
f transport *m* éolien
e transporte *m* eólico
i trasporto *m* eolico
p transporte *m* eólico
d Windtransport *m*

* **wind, up ~ → 1522**

* **wind vane → 146**

* **wind velocity → 3035**

3037 windward *adj*
f au vent
e barlovento
i sopravvento
p barlavento
d windwärts

3038 windward flood
f marée *f* de côté du vent
e marea *f* de barlovento
i marea *f* di sopravvento
p maré *f* de barlavento
d Luvflut *f*

3039 windy *adj*; **stormy** *adj*
f venteux; orageux
e ventoso; tempestuoso
i ventoso; tempestoso
p ventoso; tempestuoso
d stürmisch

3040 winter
f hiver *m*
e invierno *m*
i inverno *m*
p inverno *m*
d Winter *m*

3041 winter *v*
f mettre en hivernage; hiberner
e invernar
i invernare; preparare per l'inverno
p invernar; hibernar
d einwintern

3042 winter annual
f annuelle *f* d'hiver
e hibernal *m*; anual *m* invernal
i annuale *f* invernale
p planta *f* anual de inverno
d Winterannuell *f*

3043 winter barley
f orge *f* d'hiver
e cebada *f* de invierno
i orzo *m* invernale
p cevada *f* de inverno
d Wintergerste *f*

3044 winter catch crop
f culture *f* dérobée d'hiver
e cultivo *m* intermedio de invierno
i coltura *f* intercalare vernina
p cultura *f* intercalar de inverno
d Winterzwischenfrucht *f*

3045 winter cauliflower
f chou-brocoli *m*
e coliflor *f* de invierno
i cavolfiore *m* invernale
p couve-flor *f* de inverno
d Winterblumenkohl *m*

3046 winter characteristic
f caractère *m* hivernal
e característica *f* de invierno
i caratteristica *f* d'inverno
p característica *f* de inverno
d Wintermerkmal *n*

3047 winter corn; winter grain
f céréales *fpl* d'automne
e cereales *mpl* de invierno
i cereali *mpl* invernali
p cereais *mpl* de inverno
d Wintergetreide *n*

3048 winter cover
f couverture *f* d'hiver
e cubierta *f* invernal
i copertura *f* invernale
p cobertura *f* de inverno
d Winterdecke *f*

3049 winter crop
f récolte *f* hivernale
e cosecha *f* de invierno
i coltura *f* invernale
p cultura *f* de inverno
d Winterkultur *f*

3050 winter frost
f gel *m* d'hiver
e helada *f* de invierno
i gelo *m* invernale; gelata *f* d'inverno
p geada *f* de inverno
d Winterfrost *m*

3051 winter furrow
f labour *m* d'hiver
e surco *m* de invierno
i aratura *f* invernale
p rego *m* de inverno
d Winterfurche *f*

3052 winter garden
f jardin *m* d'hiver
e jardín *m* de invierno
i giardino *m* d'inverno
p jardim *m* de inverno
d Wintergarten *m*

*** winter grain → 3047**

3053 winter grazing
f pâturage *m* d'hiver
e pastoreo *m* de invierno
i pascolo *m* invernale
p pastoreio *m* de inverno
d Winterweide *f*

*** winter hardiness → 583**

3054 winter hardy *adj*
f résistant au froid
e resistente al invierno
i resistente al freddo
p resistente ao frio
d winterhart; winterfest

3055 wintering; in-wintering
f hivernage *m*; hivernement *m*
e invernada *f*
i invernamento *m*
p invernada *f*
d Überwinterung *f*

3056 wintering charges
f dépenses *fpl* d'hivernage
e costas *fpl* de invernada
i spese *fpl* di svernamento
p despesas *fpl* de invernada
d Überwinterungskosten *fpl*

3057 wintering harbour
f port *m* d'hivernage
e puerto *m* de invernada
i porto *m* di svernamento
p porto *m* de invernada
d Winterhafen *m*

3058 winter killing
f destruction *f* par le froid *(des semis)*
e destrucción *f* durante el invierno *(de la siembra)*
i distruzione *f* causati dal freddo invernale *(dei seminati)*
p destruição *f* pelo frio *(da sementeira)*
d Auswinterung *f*

3059 winter monsoon
f mousson *f* d'hiver
e monzón *m* de invierno
i monsone *m* di inverno
p monção *f* de inverno
d Wintermonsun *m*

3060 winter navigation
f navigation *f* d'hiver
e navegación *f* de invierno
i navigazione *f* invernale
p navegação *f* de inverno
d Winterfahrt *f*

3061 winter pruning
f taille *f* d'hiver
e poda *f* invernal
i potatura *f* invernale; potatura *f* secca
p poda *f* de inverno
d Winterschnitt *m*

3062 winter purslane
f pourpier *m* d'hiver
e verdolaga *f* de invierno
i portulaca *f* d'inverno
p beldroega *f* do inverno
d Winterportulak *m*

3063 winter quarter
f quartier *m* d'hiver
e refugio *m* de invierno
i quartiere *m* d'inverno
p quartel *m* de inverno
d Winterquartier *n*

3064 winter rest period
f repos *m* hivernal; dormance *f* hivernale
e reposo *m* invernal
i riposo *m* invernale
p dormência *f* invernal
d Winterruhe *f*

3065 winter rye
f seigle *m* d'hiver
e centeno *m* de invierno
i segale *f* invernale
p centeio *m* de inverno
d Winterroggen *m*

3066 winter savory
f sarriette *f* vivace
e ajedrea *f*
i santoreggia *f* invernale
p segurelha *f*
d Winterbohnenkraut *n*

3067 winter schedule
f horaire *m* d'hiver
e horario *m* de invierno
i orario *m* invernale
p horário *m* de inverno
d Winterplan *m*

3068 winter soil moisture
f humidité *f* hivernale
e humedad *f* invernal
i umidità *f* invernale
p umidade *f* invernal
d Winterfeuchtigkeit *f*

3069 winter solstice
f solstice *m* d'hiver
e solsticio *m* de invierno
i solstizio *m* d'inverno
p solstício *m* de inverno
d Wintersonnenwende *f*

3070 winter spraying
f pulvérisation *f* d'hiver
e pulverización *f* de invierno
i trattamento *m* invernale
p pulverização *f* de inverno
d Winterspritzung *f*

3071 winter tillage
f travaux *mpl* d'hiver
e labores *mpl* de invierno
i lavori *mpl* invernali
p colheitas *fpl* de inverno
d Winterbestellung *f*

3072 wintertime
f heure *f* d'hiver
e hora *f* de invierno
i ora *f* invernale
p hora *f* de inverno
d Winterzeit *f*

3073 winter wheat
f blé *m* d'hiver
e trigo *m* de invierno
i frumento *m* invernale
p trigo *m* de inverno
d Winterweizen *m*

3074 winter window
f fenêtre *f* d'hiver
e ventana *f* de invierno
i finestra *f* per l'inverno
p janela *f* de inverno
d Winterfenster *n*

3075 wireless bearing
f relèvement *m* radiogoniométrique
e marcación *f* radiogoniométrica
i rilevamento *m* radiogoniometrico
p marcação *f* radiogoniométrica
d Funkpeilung *f*

3076 wisp of smoke
f panache *m* de fumée
e penacho *m* de humo
i ciuffo *m* di fumo
p mecha *f* de fumo
d Rauchfahne *f*

3077 with a wind force ...
f par un vent force ...
e con viento fuerza ...
i con vento forza ...
p com vento força ...
d bei Windstärke ...

3078 wood meadowgrass
f pâturin *m* des bois
e poa *f* espiguilla
i fienarola *f* dei boschi
p capim *m* sempre verde
d Hainrispengras *n*

3079 xenon
(one of the noble and inert gases present in the atmosphere)
f xénon *m*
e xenón *m*
i xenon *m*
p xenônio *m*
d Xenon *n*

3080 xerophytes
f xérophytes *fpl*
e xerófitas *fpl*
i xerofite *fpl*
p xerófitas *fpl*
d Trockenpflanzen *fpl*

Y

3081 yard
f yard *m*
e yarda *f*
i iarda *f*
p jarda *f*
d Yard *n*

3082 year
f année *f*
e año *m*
i anno *m*
p ano *m*
d Jahr *n*

3083 yellow fever
f fièvre *f* jaune
e fiebre *f* amarilla
i febbre *f* gialla
p febre *f* amarela
d Gelbfieber *n*

3084 young ice; new ice
f glace *f* jeune
e hielo *m* reciente
i ghiaccio *m* di recente formazione
p gelo *m* recente
d Jungeis *n*

Z

* **zastruga** → 2305

3085 zenith
(the point of the celestial sphere vertically above the earth's surface)
f zénith *m*
e zenit *m*
i zenit *m*
p zênite *m*
d Zenit *m*

3086 zero
f zéro *m*
e cero *m*
i zero *m*
p zero *m*
d Null *f*

3087 zero visibility
f visibilité *f* nule
e visibilidad *f* nula
i visibilità *f* nulla
p visibilidade *f* nula
d keine Sicht *f*

3088 zodiac
f zodiaque *m*
e zodíaco *m*
i zodiaco *m*
p zodíaco *m*
d Zodiakus *m*

3089 zodiacal light
f lumière *f* zodiacale
e luz *f* zodiacal
i luce *f* zodiacale
p luz *f* zodiacal
d Zodiakallicht *n*

3090 zonda
(hot north wind blowing across the Argentina pampas)
f zonda *m*
e zonda *m*
i zonda *m*
p zonda *m*
d Zonda *m*

3091 zone of audibility
f zone *f* d'audibilité
e zona *f* de audibilidad
i zona *f* di udibilità
p zona *f* de audibilidade
d Hörbarkeitzone *f*

3092 zone of maximum auroral frequency
f zone *f* du maximum de fréquence aurorale
e zona *f* de máxima frecuencia auroral
i zona *f* di massima frequenza delle aurore
p zona *f* de máxima frequência auroral
d Zone *f* maximaler Polarlichthäufigkeit

3093 zone of silence
f zone *f* de silence
e zona *f* de silencio
i zona *f* del silenzio
p zona *f* de silêncio
d Zone *f* des Schweigens

3094 zone time
f temps *m* fuseau
e huso *m* horario
i fuso *m* orario
p fuso *m* horário
d Zonenzeit *f*

Français

Español

Italiano

Português

Deutsch